ECONOMIC ISSUES, PROBLEMS AND PERSPECTIVES

INTERNATIONAL FINANCIAL INSTITUTIONS, CLIMATE CHANGE AND THE URGENCY TO FACILITATE CLEAN ENERGY INVESTMENT IN DEVELOPING AND EMERGING MARKET ECONOMIES

Economic Issues, Problems and Perspectives

Additional books in this series can be found on Nova's website under the Series tab.

Additional e-books in this series can be found on Nova's website under the eBook tab.

Financial Institutions and Services

Additional books in this series can be found on Nova's website under the Series tab.

Additional e-books in this series can be found on Nova's website under the eBook tab.

ECONOMIC ISSUES, PROBLEMS AND PERSPECTIVES

INTERNATIONAL FINANCIAL INSTITUTIONS, CLIMATE CHANGE AND THE URGENCY TO FACILITATE CLEAN ENERGY INVESTMENT IN DEVELOPING AND EMERGING MARKET ECONOMIES

HILMAR ÞÓR HILMARSSON, PhD

nova publishers
New York

Library of Congress Cataloging-in-Publication Data

ISBN: 978-1-53610-045-7

Published by Nova Science Publishers, Inc. † New York

To my mother, Hólmfríður Snorradóttir

CONTENTS

PREFACE

This book has an international focus – the global challenges of climate change and environmental degradation – and the urgency to take action and scale up funding for clean energy projects in developing and emerging markets to help battle these challenges.

The coming years and decades will inevitably see huge investments in global infrastructure. In a recent speech, the World Bank Group president, speaking about climate change, stated that in the next two decades more investment will be poured into infrastructure throughout the world than in the preceding 6000 years.[1] This is staggering, and even if it is an overestimate, the scale of new infrastructure investment in the near future will be enormous, especially in developing and emerging market economies. It is important that this new infrastructure be climate friendly and that environmental considerations are given the attention they deserve. The right choices have to be made and part of that strategy will be to support and promote clean energy investment by increasing the use of clean energy sources that are available on earth.

Given the international focus of this book, the reader may be surprised to learn that my initial thinking on the topic began when observing changes in the energy mix of my own country, Iceland. In a few decades Iceland has made a unique transition to clean energy, using primarily geothermal energy and hydropower. Utilization of clean energy for electricity production and space heating has been beneficial for the nation for decades and has also been critical

---

[1] Speech by World Bank Group President Jim Yong Kim at Georgetown University on 19 March 2015. https://www.georgetown.edu/news/world-bank-group-president-on-climate-change.html.

for Iceland's transition from a developing country to a high income developed country.

Geothermal resources can be divided into two segments: low temperature for direct use of heat, and high temperature for indirect use e.g., electricity production. Among the direct use activities are district heating, as well as multiple use activities such as agriculture, fishing, tourism, bathing, recreation and health. In Iceland, this means, for example:

- district heating: heat for homes and organizations, soil heating (snow melting) for public and private places.
- agriculture and fishery; greenhouses, fish drying and fish farms.
- tourism, bathing, recreation and health: among more than 130 swimming pools in a country of about 330, 000 people, the world famous Blue Lagoon and Mývatn Nature Baths.

Additionally, indirect use (electricity) operations come to mind, such as aluminum smelters, data centers, seaweed manufacture, and other industrial activities. Considering the benefits of geothermal energy thus involves considering multiple-use activities that all generate income and increase the well-being of the nation and its visitors. These benefits come in addition to the critical local and global benefit of moving from fossil fuel to clean energy sources. With that in mind I wondered if a small country like Iceland could share its experience over recent decades to make meaningful contributions on a global scale. Could developing countries and emerging market economies learn from this experience in their quest for growth and at the same time develop in a way that is environmentally friendly for themselves as well as globally?

We live in an era when a growing number of scientists believe that climate change is related to human activity and is a major threat to the existence and well-being of mankind. At the same time, the existence of a relatively small group of skeptics among scientists must also be noted. But even if climate change as a problem was not as serious, as most scientists now believe, the environmental degradation that comes with increased use of fossil fuel is clear beyond doubt. Utilization of clean energy sources can also generate multiple use activities, as noted above, thus increasing people's income and well-being in an environmentally friendly way. I wondered if Iceland might make contributions that could help make the world a better place, contributions beneficial for Iceland herself and for the world as well. Can a small country like Iceland help others while helping herself? If not, then why not? If yes, then how?

When considering these questions, I soon came to the conclusion that Iceland could be a source of inspiration and encouragement for other nations, especially on geothermal development. Many countries have successfully utilized hydropower as Iceland has done, but in geothermal utilization Iceland is unique: Iceland's utilization of geothermal energy can show the world what is possible. Iceland can also share its experience. But Iceland is too small to make important contributions on a global scale unless it partners with other countries and institutions, especially global institutions.

Nevertheless, Iceland is undoubtedly a significant player in the global geothermal market, with the highest share worldwide of geothermal energy in its overall energy use. At the same time, Iceland possesses both strengths and weaknesses in this area. Among the strengths are highly experienced companies, institutions and employees; a well-developed system for using geothermal energy in multiple ways; as well as a strong international reputation and a considerable international network. But the list of weaknesses is also long. Geothermal companies are small on a global scale, while the largest among the small are in public ownership; research and educational activities suffer from small size and fragmentation; formal platforms for collaboration are lacking despite the existence of a small geothermal cluster management group, Iceland geothermal, that has capable staff but lacks the capacity to coordinate major geothermal activities on an international scale; finally, and perhaps the most serious impediment, is access to capital.

Iceland cannot provide capital on a large scale to fund geothermal activities overseas. Its capital contributions would in most cases amount to a mere token. Moreover, Iceland has neither a well-developed network nor the capacity to take the lead in forming international consortia in cooperation with major funding sources, including international financial institutions and global investment banks.

Geothermal energy offers serious potential globally in the medium and long term. Currently, however, geothermal is only a niche market compared to hydro, solar and wind. Nevertheless, geothermal energy presents an interesting global market opportunity, with significant growth rates in the coming years, bearing in mind the fact that only about 5% of potential capacity has so far been exploited.

We know that most clean energy resources, including geothermal, are located in developing and emerging market economies (for example, about 40 percent of unexploited global geothermal capacity is located in Asia). This is also where most of the future economic growth, and thus also growth in energy demand, is likely to take place. We know, too, that clean energy investments tend to be large, capital intensive and with long repayment periods. But capital is scarce in many of these countries, so capital mobilization is important from all sources, including the private sector. As noted above, it is important that new infrastructure should be climate- and environmentally friendly and at the same time increase the welfare of the nations concerned via multiple use activities. In particular, multiple use activities that increase value added can help developing countries move from low income country status to middle income country status and from middle income countries to high income ones.

In addition to capital mobilization challenges, geothermal projects also suffer from high upfront costs and resource risks. This in turn makes their financing risks high. Thus it is difficult to fund the initial costs necessary for resource validation.

If Iceland wants to cooperate with others in geothermal development on a global scale, whom should it partner with? Potential partners in developing and emerging markets that come to mind are international financial institutions (IFIs). When speaking about IFIs I mean not only the Bretton Woods[2] institutions, in this case the International Bank for Reconstruction and Development (which later became the World Bank) but also the regional development banks including the Inter-American Development Bank, the African Development Bank, the Asian Development Bank, the European Bank for Reconstruction and Development, and so on. The list might also in the future include the more newly established IFIs: the Asian Infrastructure Investment Bank and the New Development Bank – BRICS.[3]

I began considering the instruments that IFIs employ to support clean energy projects in developing and emerging markets. I soon came to the conclusion that IFIs are in a unique position to address the challenges related to geothermal projects. IFIs support both government and private sector projects, as well as cooperating with bilateral and other multilateral institutions. They have relatively strong coordination capacity and have various lending and equity instruments to participate in project finance. IFIs also have a menu of guarantee

[2] In July 1944 delegates from 44 nations assembled at Bretton Woods in New Hampshire USA to create a new post-war global arrangement. During only a few weeks they developed a system of rules, institutions, and procedures for financial and commercial relations for the world economy. Two key institutions established at Bretton Woods still exist today: the World Bank and the International Monetary Fund. Virtually every country on earth is now a member of both institutions. The role of these twin pillars, which still support the structure of the world's economic and financial order, has evolved over time. The third pillar was the General Agreement on Tariffs and Trade, which later became the World Trade Organization.

[3] BRICS is an association of five major emerging market economies and formed by the capital letter of each: Brazil, Russia, India, China and South Africa.

and insurance instruments to mobilize funding in partnership with the private sector. Furthermore, they often engage in policy dialogue with governments. This in turn can lead to policy reforms for the whole economy, including reforms that can be critical in strengthening energy sector development. IFIs can also provide the much needed technical assistance that is often required in developing countries and emerging market economies.

Unfortunately, the government of Iceland has not invested much in human resources to build capacity to cooperate with IFIs, nor indeed with international organizations at all for that matter. Multilateral cooperation has only been given low priority within the Icelandic government machinery. Indeed, in this regard the government of Iceland has been inward looking, with multilateral contributions usually kept to a minimum. Iceland is a member of only a few IFIs; the World Bank Group, the European Bank for Reconstruction and Development, the Nordic Investment Bank and most recently the Asian Infrastructure Investment Bank.

Iceland has little capacity for working with the institutions that it is a member of, nor is Iceland a member of key regional development banks such as the African Development Bank, the Asian Development Bank and the Inter-American Development Bank. This limited capacity and engagement makes it difficult for Iceland to contribute meaningfully to geothermal development in cooperation with international institutions. Nevertheless, the country can boast many companies and institutions with technical knowledge in exploring geothermal sites as well as in building and operating geothermal power plants. Less experience is available in funding and risk mitigation for geothermal projects and cooperation with the private sector, perhaps because geothermal development in Iceland has mostly involved government support and funding.

I first criticized the Icelandic government's lack of cooperation with IFIs in a public presentation at the University of Iceland in 2008.[4] Since then Iceland has initiated cooperation with the World Bank on geothermal development, mainly in Africa with its support for the Global Geothermal Development Plan, whose purpose is to mobilize funds for early-stage investment in geothermal energy projects in developing countries. Recently the government of Iceland also decided to be among the founding members of the newly established Asian Infrastructure Investment Bank. It appears the government did not carry out any feasibility study before deciding to apply for membership in the Asian Infrastructure Investment Bank. Moreover, an overall strategy and vision for its participation in international financial institutions is still lacking, while

[4] See further: http://www.irpa.is/article/view/986.

investment in coordination capacity remains low. Supporting geothermal development in Africa and then becoming a member of the Asian Infrastructure Investment Bank raises questions about Iceland's intentions. Does it want to participate in geothermal development in Africa or Asia? Spreading small resources thin does not look like a promising strategy.

The Icelandic Ministry for Foreign Affairs conducts the relationship with the World Bank Group but strangely the Ministry of Finance and Economic Affairs conducts the relationship with the European Bank for Reconstruction and Development, the Nordic Investment Bank and will also do so for the Asian Infrastructure Investment Bank. It is likely that the Icelandic government will neglect the Asian Infrastructure Investment Bank just as much as it has neglected its participation in other IFIs it is a member of. When writing this book, I have repeatedly sent emails to the Ministry of Finance and Economic Affairs asking for details of progress relating to Iceland's preparation for membership in the Asian Infrastructure Investment Bank. It took several months just to be informed who was working on Asian Infrastructure Investment Bank issues at the Ministry of Finance and Economic Affairs. Unfortunately, this is typical of the government´s behavior in terms of public communication, including with academia. In this regard Iceland remains an underdeveloped country where accountability within the government is blurred. This became very clear during the 2008 financial crisis, when no one in the government felt responsible. Government bureaucrats in one institution pointed fingers at colleagues in other institutions.

When considering available information on clean energy investment in developing and emerging countries in recent decades, it seemed to me that IFIs hold a very modest clean energy portfolio on a global scale and appear to be risk averse. I have not yet managed to obtain a comprehensive overview of clean energy projects supported by the World Bank Group. Even if it exists, it may be too embarrassingly small to make public. However, I did manage to acquire a list of geothermal projects that the World Bank has supported as well as numbers for total engagements of the World Bank and the regional development banks i.e., the African Development Bank, the Asian Development Bank and the Inter-American Development Bank. These numbers also included the European Investment Bank. If one considers geothermal engagement, according to World Bank sources, total investment in geothermal development by the World Bank, the regional development banks mentioned above and the European Investment Bank has been about US$3 billion over the last three decades. This amounts to less than a US$1 contribution to geothermal development per capita in emerging and developing countries over a thirty-year period: only a few cents per capita per year. This modest engagement was a surprise to me. And even this tiny engagement was mostly (94 percent) allocated to the least risky part of geothermal development, i.e., when projects have already become bankable.

This raises questions. Do IFIs know something about climate change that we do not? US$3 billion is a large amount of money for an individual or a small/medium sized enterprise but in a global context it is a contribution of little significance given a world population of some 7 billion people with more than half residing in developing and emerging countries. This is also true of contributions by IFIs to other clean energy investments. With this lack of scale in contributions, IFIs are almost irrelevant in the fight against climate change in spite of all the talk and all the colorful reports, with their fancy pictures and graphs. Problems related to climate change not only affect individual countries locally, nor do they affect only developing and emerging market countries. Climate change has global consequences affecting all countries. This deserves serious attention from IFIs.

I am not unfamiliar with IFIs and international development cooperation, having worked for the World Bank for 12 years, first at its headquarters in Washington DC, mainly on the Africa region, then at its office in Riga in Latvia, and finally, at its Hanoi office in Vietnam. I have thus been involved in Bank operations on three different continents. I think this gave me a global perspective and enriched my life, though it was challenging and at times undoubtedly confusing.

During those years at the World Bank, coordination with both bilateral donors and multilateral institutions was an important component of my work. This was not easy with the love-hate relationship that sometimes exists between bilateral donors and the bank. In addition, tensions often arise between bilateral donors as well as within the World Bank, e.g., between local staff in the field and expatriates who are better paid and receive more generous benefits.

Between World Bank assignments I was the Special Advisor to the Minister for Foreign Affairs in Iceland from 1995 to 1999. This also included chairing the Icelandic International Development Agency.[5] I was appointed by the Minister for Foreign Affairs but the rest of the committee was elected by the Icelandic parliament, the *Althingi*. I also chaired a committee appointed by the Icelandic Cabinet, which was responsible for designing and implementing Iceland's emerging assistance to Bosnia and Herzegovina. I was also responsible for the ministry's coordination with the Nordic countries and the

[5] The Agency was responsible for all Icelandic bilateral development assistance until recently when the agency was merged with the Ministry for Foreign Affairs.

Baltic States on World Bank affairs. This included preparing the minister's speeches when he represented the Nordic countries and the Baltic States for the World Bank and International Monetary Fund Annual Meetings. In addition, I was responsible for preparing the Nordic and Baltic statements for the Development Committee when the Icelandic minister represented those countries at its meetings. This was very challenging, since nobody within the Ministry of Foreign Affairs was familiar with World Bank operations.

Since becoming a professor at the University of Akureyri in Iceland in 2006, I have visited the World Bank regularly and engaged in dialogue with its staff. I have lectured about development cooperation and international financial institutions in over 30 universities and talked to thousands of students. During my time at the World Bank and the Ministry of Foreign Affairs I travelled to about sixty countries.

I have enormous respect for the quality of staff at the World Bank, their expertise and hard work. One of the main strengths of the World Bank is its staff, most of whom, I think, are genuinely interested in development and want to serve the poor and disadvantaged. They are, however, often working with very limited resources compared to the task before them, often under pressure, battling a huge bureaucracy with an endless list of safeguards. It is hard to get things done in this environment, as will be discussed in this book.

***

Greater utilization of clean energy resources is important for the whole world. As noted above, concern is growing about climate change and environmental degradation. While the debate continues about the causes of climate change, growing scientific consensus suggests that human activity is accelerating this process and that climate change presents serious global risks. However, no one can predict the consequences of climate change with certainty. Personally I cannot make a judgment on this as the causes and consequences of climate change fall outside my area of expertise. My training is not in the natural sciences but in economics and public administration. Nevertheless, it would clearly be irresponsible not to take action given the mounting scientific evidence. In addition, and irrespective of climate change issues, clean energy projects can reduce environmental degradation and can dramatically improve people's quality of life.

The foundations of the World Bank were laid at the Bretton Woods conference when World War II was still ongoing but nearing its conclusion. Initially the World Bank was meant to be engaged in post war reconstruction.

The World Bank soon shifted its focus to development issues and has now become the largest and most powerful development organization in the world. At the same time, it has become complex and bureaucratic. With all its bureaucratic processes and safeguards the World Bank can rarely make an immediate impact, even under streamlined procedures during times of crisis. In contrast the International Monetary Fund, the World Bank's sister institution created at Bretton Woods, is able to respond immediately to financial crises. I remember clearly an interview that I took in Washington in 2010 with a distinguished World Bank staff member who had been a country director. When we talked about the World Bank and its ability to cooperate with the private sector to mobilize funding for a major clean energy project in Asia he said: "The bad news is that the World Bank is seen as a high-cost/high-hassle partner of last resort. There is therefore a critical need to reduce the costs the private sector incurs for doing business with the World Bank. Doing so will require the World Bank to better understand the constraints under which the private sector works."[6]

One should also not forget that the World Bank Group is owned and run primarily by rich countries. Most influential are the G7[7] countries, primarily the US. Why? As noted above, the Bank was established during the Bretton Woods conference in the United States (USA/US) during a time when World War II was ending and the USA was the dominant world power with the strongest economy. The president of the World Bank is traditionally appointed by the president of the United States. The World Bank's headquarters are located in Washington DC. The US executive director representing the US on the World Bank's board can walk over to the US treasury, attend high level meetings, and bring the message from those meetings to the World Bank's president and its board. This, in addition to the fact that the US remains the most powerful country in the world, gives the US enormous power at the World Bank, beyond the size of its economy and its voting power within the World Bank.

The world is now changing. The fast growing economies of key emerging market countries have made G20[8] more important and powerful than before, so that today G20 is challenging the dominance of the G7 powers. The

[6] Interview at the World Bank June 25, 2010. See further Porter, I. C. and Shivakumar, J. (2010).

[7] The G7 are the seven major advanced economies. In addition to the United States, these are: Canada, France, Germany, Italy, Japan, and the United Kingdom. The European Union is also represented within the G7.

[8] The G20 is an international forum for 20 major economies. The members include 19 countries: Argentina, Australia, Brazil, Canada, China, France, Germany, India, Indonesia, Italy, Japan, South Korea, Mexico, Russia, Saudi Arabia, South Africa, Turkey, the United Kingdom and the United States -along with the European Union.

establishment of the Asian Infrastructure Investment Bank and the New Development (BRICS) Banks, both led by China, challenges the Bretton Woods institutional establishment as never before. The World Bank's role in the world and its US dominance is now contested and it appears to me that the World Bank needs to change. If establishment of the Asian Infrastructure Investment Bank, led by China, scales up clean energy investment, this development should be welcomed. Even so, global climate change efforts will not succeed without US participation.

Global needs for investment in clean energy are vast and can only be met if resources are pooled from different sources; public, private, and donor funds. While government funds are important, many developing and emerging countries do not have the tax base to fund large, capital intensive, long term projects. Effective and efficient International Financial Institutions could help capital mobilization for clean energy projects and facilitate cooperation with the private sector. This calls for public-private partnerships (PPPs),[9] which can be challenging in developing and emerging countries, as this book will discuss in some detail. Furthermore, private resources will only flow into developing and emerging countries to scale up climate friendly projects if risks can be managed. Here the IFIs can play a key role. However, it is also important that, when the private sector engages in these countries, the deal is fair for all stakeholders, including the host country. Allocation of rewards and risks must be fair and sustainable for the host country, private investors and lenders, as well as multilateral and bilateral institutions involved. The private sector should not demand excessive profits when offered reasonable risk mitigation by the international community via IFIs and export credit agencies.

One important obstacle in scaling up clean energy projects today is that no comprehensive multilateral agreement on foreign investments exists. Additionally, no multilateral institution is engaged in cross border investments in the same way as for cross border trade. No World Trade Organization exists for investment. This can be an obstacle for global efforts to promote cross border climate change investment and an impediment for private sector flows.

Regarding geothermal energy, technical and engineering knowledge is growing on how to utilize this resource effectively and efficiently. However, a lack of expertise persists in project development and management. More experts are needed to develop business plans, to carry out necessary financial analysis and design suitable instruments to promote the sector. More progress in utilizing geothermal energy requires a multidisciplinary approach. Utilization of clean

[9] See Chapter 7 for discussion on definitions of public-private partnerships.

energy resources to address climate change issues is no longer an engineering problem only. It has become an economic, financial and political issue that requires the attention and the cooperation of many different experts.

Lack of enthusiasm is rife among the public, who do not rank the transition to clean energy as a high priority issue, or among politicians, who do not see the urgency for action. Countries will also respond to these challenges differently under different circumstances; for example, China with a single party government and the USA with its democratic processes. In addition, well organized, well-funded and powerful industries that produce or use fossil fuels do not see clean energy investment as a priority. They represent a small minority but their influence on elected officials is strong in many countries.

While investment in clean energy, including geothermal, is important to protect the environment and fight climate change, the author does not suggest that this alone is sufficient to solve the problem. Policies to eliminate global warming externality[10] may also require an appropriate price for carbon. Clean energy investment and a tax on carbon should be explored simultaneously. If price policies fail, clean energy investment may be our last hope for achieving our climate goals. Incentivizing emerging investors to prefer low-carbon investment is an important step in the right direction that the world cannot afford to neglect.

[10] Externality is the cost or benefit of one person's decision on the well-being of a bystander (a third party) which the decision maker does not take into account in making a decision. The consequences of global warming are a negative externality i.e., the costs imposed on a third party by a decision.

ACKNOWLEDGMENTS

While writing this book I have benefited from visiting appointments, most recently at Cornell University during the Fall semester 2015, as well as at the University of Washington during Spring 2014 and the Stockholm School of Economics in Riga in the Fall semester 2013. During the Fall semester 2015 I benefited from meeting with experts and discussing issues related to climate change at universities in the USA, including Cornell University, Harvard University, the Massachusetts Institute of Technology (MIT), Princeton University and Tufts University. I also benefited from meetings and information provided by staff of the World Bank Group.

When writing this book, I benefited especially from communicating with: Jeffrey Delmon, Trung Quang Dinh, Þráinn Eggertsson, Ravi Kanbur, Peter J. Katzenstein, Robert O. Keohane, Michal C. Moore, Jeswald W. Salacuse, Jayasankar Shivakumar, Jefferson William Tester, Louis T. Wells and Gerald T. West.

Finally, I would like to think my language editor Christopher Goddard for a careful review and excellent editing of the text.

ACRONYMS AND ABBREVIATIONS

ASEAN	Association of Southeast Asian Nations
BOT	Build-Operate-Transfer
BRICS	An acronym for an association of five major emerging market economies: Brazil, Russia, India, China and South Africa
COFACE	French company specializing in export credit insurance
EBRD	European Bank for Reconstruction and Development
EIB	European Investment Bank
EU	European Union
FMO	Dutch development bank
G7	The Group of Seven: are the seven major advanced economies. In addition to the United States, these are: Canada, France, Germany, Italy, Japan, and the United Kingdom. The European Union is also represented within the G7.
G20	The Group of Twenty: an international forum for 20 major economies. The members include 19 countries: Argentina, Australia, Brazil, Canada, China, France, Germany, India, Indonesia, Italy, Japan, South Korea, Mexico, Russia, Saudi Arabia, South Africa, Turkey, the United Kingdom and the United States - along with the European Union.
IBRD	International Bank for Reconstruction and Development
ICSID	International Center for Settlement of Investment Disputes
IDA	International Development Association
IFC	International Finance Corporation
IFI	International financial institutions
LIBOR	London Interbank Offered Rate
MIGA	Multilateral Investment Guarantee Agency

NTPC	Nam Theun 2 Power Company Limited
NT2	Nam Theun 2
PPIAF	Public-Private Infrastructure Advisory Facility
PPP	public-private partnership
SREP	Scaling-Up Renewable Energy in Low Income Countries Program
TRÚ	Tryggingardeild útflutnings (Export Credit Agency, Iceland)
UN	United Nations
US/USA	United States/of America
US$	United States dollar (currency)

SUMMARY

Climate change is perhaps among the most serious challenges humankind has ever faced and maybe the greatest market failure the world has ever seen. Although skeptics can still be found, the scientific consensus of the seriousness of the climate threat has increased in recent years with a growing number of scientists believing that climate change is related to human activity. Climate change may seriously affect nations both rich and poor in the not so distant future. But, even if climate change was not a problem as serious as most scientists now believe, the environmental degradation that comes with continued and increased use of fossil fuel is clear beyond doubt. It reduces the quality of life for humans, including people's health. Use of fossil fuel not only reduces the quality of life in the country where it is used. It also reduces the quality of life for people in other countries. Indeed, what happens in one country does affect other countries. This is thus a global problem.

At the same time clean energy resources are available around the world that, if properly utilized, could help remedy some of these problems. One can argue that these resources are to a large extent in the "wrong" places to be exploited efficiently and effectively, i.e., in developing and emerging market economies that are often in economic and political transition with limited capacity to take action and to carry out the necessary exploration to make projects bankable. Some problems are manmade obstacles.

It is likely that most of the increased demand for energy in the future will be in the developing and emerging world, which is currently enjoying relatively high economic growth but has not yet reached high income levels. Clean energy sources are also to a large extent located in these countries but remain underutilized. Geothermal resources are a good example of a clean energy source that is underutilized in many regions of the emerging world.

The challenge of climate change requires a strong comprehensive response including investment for use of more renewable resources and less fossil fuel. However, the international investment regime is divided and no international organization, or government, is available to address these environmental and climate challenges holistically on a global scale. The absence of a global regime makes climate change an acutely difficult international problem to manage.

Currently, strong incentives exist for individual countries to free ride. As discussed in this book, rich countries have preferred to sign bilateral investment treaties with developing and emerging countries rather than create a global investment regime supported by an international organization in the same way as the global trade regime is supported by the World Trade Organization. This makes long term clean energy investment in developing and emerging markets more problematic. International Financial Institutions (IFIs) also tend to serve investors from rich countries rather than the interests of developing host countries. Self-interest and short sighted (in)action by rich countries make climate change a prickly problem to manage.

Among the remedies for addressing climate change challenges feature reduction in the use of fossil fuel and utilization of more of the earth's clean energy resources. This requires major long term investment in infrastructure. It is clear that significant investment in infrastructure will take place in the decades to come. Critically, those investments should be climate friendly and, to as large an extent as possible, promote the use of clean energy resources.

To meet such vast infrastructure needs requires the utilization of private, public and donor resources, especially in developing and emerging economies where capital is scarce. However, funds for clean energy resources are not flowing cross border into those countries in line with their needs, often due to challenging business and investment climates. These may be manmade obstacles, including political or non-commercial risks.

To mitigate risks, IFIs offer piecemeal solutions that so far have not been used on a large scale. These institutions have hesitated to use their guarantee powers to mobilize clean energy funding. While IFIs respond swiftly to financial crises, their reaction to the climate crisis is slow, almost as if they knew something that climate scientists do not.

The partners typically involved in infrastructure investment may also not work in harmony so as to result in favorable global outcomes. For example, international institutions that are supposed to coordinate global action also have their own agenda and can be ineffective in addressing the issues they were created to solve. They may be more concerned about their net income and a healthy financial structure, resulting in a strong credit rating, than the global good. This can make them risk averse to the extent that they do not respond effectively to global problems. The rich countries that own the IFIs to a large extent can change this. However, so far political will is lacking.

Private companies that invest in risky markets may demand quick and "excessive" returns even if risk mitigation is offered by the international community via IFIs. They may want to grab whatever profits they can as quickly as possible and then exit the country. This does not sit well with clean energy projects, which tend to require long term commitment. Private investors may walk away from projects in developing and emerging markets instead of renegotiating during times of crisis. Host governments may have agendas other than maximizing the welfare of their people. Government in developing and emerging markets with limited institutional capacity may also be vulnerable when dealing with large multinational companies as well as IFIs. To be successful and sustainable these projects need to be approached and executed fairly and efficiently. The deal needs to be fair for all the partners in terms of allocation of both risks and rewards.

The fossil fuel industry has strong incentives to ensure that this source continues to be used. At the same time the public in many countries may consider other problems more pressing than climate change and environmental degradation. Thus a small group of strong, well organized private interests gets its way while a large group of people who will be affected but is less organized suffers the consequences.

Clean energy projects tend to be large, capital intensive and long term. They require long term commitment from all the players involved as well as mutual trust. This book focuses on those challenges, mainly using geothermal energy projects as examples. It also provides examples from large hydropower projects to illustrate how the funding and risk mitigation instruments of IFIs have been used to mobilize funds in difficult investment environments.

Finally, a note for the reader: When reading the book cover to cover the reader will notice a certain amount of repetition. This is because each chapter is meant to some extent to be independent for those who prefer to dip into the book to study specific issues instead of reading the whole book in traditional "start-to-finish" style.

INTRODUCTION

CLEAN ENERGY, ENVIRONMENTAL DEGRADATION AND CLIMATE CHANGE – THE GLOBAL CHALLENGE

Climate change is perhaps among the most serious challenges humankind has ever faced. This is a global problem that has the potential to become a fundamental threat to economic development and human well-being on the earth, so that urgent action is needed.

Natural scientists are not the only ones expressing concern about climate change. Economists, financial analysts and political scientists, among others, are now also increasingly engaged in this field. The problem is no longer only one for geoscientists and engineers. Climate change can have serious economic, financial and political consequences. Sir Nicholas Stern, a former chief economist at the World Bank Group, led a panel to review the evidence for climate change and to assess its economic implications. The Stern Commission's key message was that: "An overwhelming body of scientific evidence now clearly indicates that climate change is a serious and urgent issue. The Earth's climate is rapidly changing, mainly as a result of increases in greenhouse gases caused by human activities." (Stern Commission, 2006, p. 2). The Intergovernmental Panel on Climate Change states that "The human influence on the climate system is clear and is evident from the increasing greenhouse gas concentrations in the atmosphere, positive radiative forcing, observed warming, and understanding of the climate system."[11] In his book *Earth – A Tenant's Manual*, Rhodes states that "These increases in greenhouse gases are the clear result of human activity, reflecting the growing emissions

[11] See Intergovernmental Panel on Climate Change, Fifth Assessment Report, see https://www.ipcc.ch/report/ar5/index.shtml.

produced by increased industrialization, transport, deforestation, intensive agriculture, urbanization, and growing population. And the gases have long atmospheric lifetimes…." (Rhodes, 2012, p. 172). Political scientists are also turning their attention to the problem. Robert Keohane, who has published extensively on the global politics of climate change, states that scientific consensus on the seriousness of the climate threat has increased (Keohane, 2015, p. 20). Finally, the World Bank recently stated that "science is unequivocal that humans are the cause of global warming, and that major changes are already being observed" (World Bank, 2015a, p. 18).

Although the consensus on the seriousness of the climate threat has increased in recent years, some debate still remains about the influence of human activity on climate change. Rhodes, for example, states that: "Because records of high greenhouse gas concentrations show [a] strong correlation with higher temperatures in ice cores, these increases are assumed to compound natural temperature increases. There is no significant disagreement in the scientific community on this point. The debate, rather, is about the degree of impact and future scale and timing of human influence on global warming" (Rhodes, 2012, p. 172). A small number of skeptics can still be found among natural scientists about the causes of climate change but this book will not discuss the matter in any detail.

The consequences of climate change can become dramatic. In his book, The Climate Casino – Risk, Uncertainty, and Economics for a Warming World, Nordhaus, for example, states that these include sea-level rise, more intense hurricanes, losses of species and ecosystems, acidification of the ocean, as well as threats to the natural and cultural heritage of the planet (Nordhaus, 2013). Currently the most striking evidence of climate change perhaps comes from the shrinking coverage of polar ice as can be seen in Greenland and Iceland, for example.

When discussing the effective response to climate change, the Stern Review states that: "Climate change is the greatest market failure the world has ever seen, and it interacts with other market imperfections. Three elements of policy are required for an effective global response. The first is the pricing of carbon, implemented through tax, trading or regulation. The second is policy to support innovation and the deployment of low-carbon technologies. And the third is action to remove barriers to energy efficiency, and to inform, educate and persuade individuals about what they can do to respond to climate change" (Stern Review, 2006, p. viii). While all those issues are important, this book will mainly focus on the deployment of low carbon technologies via the utilization of underutilized renewable energy sources.

In economic theory pollution is considered a negative externality.[12] It is a by-product of human activity and pollutants cause damage to innocent bystanders. Emission of carbon dioxide and other greenhouse gases is causing damage now and will do so in the future. These gases negatively impact global climate change. As Nordhaus put it "The problem is that those who produce the emissions do not pay for that privilege, and those who are harmed are not compensated" (Nordhaus, 2013, p. 17) and "governments must step in and regulate or tax activities with significant harmful externalities." (Nordhaus, 2013, p. 19)

One can distinguish between local and global externalities. Emission of carbon dioxide can for example adversely impact the health of people in the country they live in and can thus be called negative local externality. But carbon dioxide emissions not only affect local communities. They also have a negative impact on the entire global community that stands to lose from increased carbon dioxide emissions. This is because carbon dioxide emissions from one country affect all countries through their impact on global climate. Human activity thus imposes long lasting costs on bystanders without compensation. This is the essence of the market failure.

Any solution to the problem of climate change must be a global solution. Piecemeal solutions will not do. The participation and leadership of all major powers, most notably the United States, the European Union (EU), China, and India is required. As Frank Rhodes states: "The United States…has about 5 percent of the world's population but produces about a quarter of all [greenhouse gas emissions]. Though China now slightly exceeds the U.S. total emission, it has a population of four times that of the United States, and its level of [greenhouse gas emissions] is growing rapidly" (Rhodes, 2012, p. 181). The battle against climate change thus requires cooperation and coordination among developing, emerging and high income countries.

This book focuses on investment in clean renewable energy sources, mainly geothermal. This is because investment in renewable energy sources can help shift the world away from high-carbon fossil fuels. Its focus is on investment in developing and emerging markets, currently the fastest growing economies in the world in terms of Gross Domestic Product growth, energy use and emissions. Most of the future increase in energy demand will be in developing and emerging markets. This is also where most clean energy sources are located. In addition to environmental benefits, multiple uses of clean energy sources

[12] Externality can be defined as a consequence of an industrial or commercial activity which affects other parties without this being reflected in market prices, see for example: http://www.oxforddictionaries.com/definition/english/externality

such as geothermal can also increase value added in developing and emerging countries and help them in their transition from low to middle income, and eventually to high income countries.

It is widely believed that global warming will have more serious consequences in developing countries than in high income developed countries. We know that developing countries currently possess vast underutilized clean energy resources. Why are these resources not used on a larger scale and what needs to be done to take advantage of those resources to help remedy the problems of climate change for those countries and, indeed, for the whole world?

Many challenges are related to increased use of renewable sources such as geothermal and hydropower. As the World Bank has stated: "Capital intensive infrastructure projects have a number of distinctive features: (i) they require significant upfront capital and take many years to payback; (ii) output is typically sold on a basis of long term contracts; (iii) and permitting risks can be significant" (World Bank, 2012a, p. 3). And furthermore: "low-emission projects tend to have higher upfront costs; produce less output per unit of capacity; and have higher perceived risks than conventional infrastructure projects."

The geothermal sector is currently a market niche that has a substantial global market opportunity with significant growth rates in coming years. In fact, according to the International Financial Corporation: "Around 40 countries worldwide, including several low and middle income countries, have the potential to meet a sizeable proportion of their electricity demand through geothermal power, at a relatively low cost (around $.08 per kilowatt hour)" (International Financial Corporation, 2014, p. 52). One can say that geothermal is like an infant that has not been taken care of. The need is to step up local and global efforts. This book will also discuss large hydropower projects built in a challenging business and investment environment to show how the international community can help make such investments available with the appropriate lending and risk mitigation instruments. This requires the cooperation, coordination and commitment of many different players, including host governments, the international community via international financial institutions, bilateral agencies and donors as well as the private sector. The World Bank recently stated that it "is working to leverage both public and private sources of climate finance to support climate-smart policy and investments and help countries and business adapt to a changing climate" (World Bank, 2015a, p. 18).

Any action to make a transition to clean energy is likely to be met with resistance. A small minority of well represented industries with large funds are likely to block policies that would benefit the larger, long-term interests of the majority. Those industries, largely concentrated in sectors that produce or use fossil fuels, will feel economically disadvantaged. No one should underestimate the power of the coal and oil industries. Change is likely to encounter stubborn resistance from those industries.

The global warming externality can be eliminated by setting an appropriate price for carbon. But further action is needed. Indeed, simultaneous multiple action is required, including: (i) the pricing of carbon (implemented through tax, trading or regulation), (ii) policy to support innovation and the deployment of low-carbon technologies, and (iii) action to remove barriers to energy efficiency, and to inform, educate and persuade individuals about what they can do to respond to climate change.

This includes clean energy investments, which are the main topic of this book. As Nordhaus has argued, "…if other policies fail, development of low-carbon technologies is the last refuge for achieving our climate goals" (Nordhaus, 2013, p. 289). Currently, renewables account for only a small share of global primary energy consumption, which is still dominated by fossil fuels: about 30 percent each for coal and oil and 25 percent for natural gas (see Table 1). Progress in the development of renewables could be fragile if fossil fuel prices remain low for long. Low prices for oil, gas and coal can slow down innovation and adoption of cleaner technology. Policymakers should not allow this to happen.

Table 1. Share of Primary Energy Consumption in 2014 (percentage points)

	Oil*	Natural Gas**	Coal**	Nuclear Energy	Renewables
World	33	24	30	4	9
United States	36	30	20	8	5
China	18	6	66	1	10
European Union	37	22	17	12	13

*Oil is mostly to power transportation

**Coal and natural gas are mainly used for electricity generation

Source: International Monetary Fund, 2015[13]

[13] The Price of Oil and the Price of Carbon, see http://blog-imfdirect.imf.org/2015/12/02/the-price-of-oil-and-the-price-of-carbon/.

The 2016 Paris climate summit[14] adopted a new agreement. It does not ensure or spell out the end of fossil fuels or assure that temperatures will not rise more than two degrees. Nevertheless, the Paris Agreement has the potential to mark a historic shift in how the world negotiates cooperation on climate change. Only time will tell if the full promise of the Paris Agreement will ever be achieved.

Every five years, beginning in 2020, each country will be expected to contribute a new national plan for reducing emissions. The agreement includes a promise to grow financial support for poorer countries beyond US$100 billion after 2020. These funds are to assist poor countries in transitioning to low-carbon economies. Part of this effort is to increase the use of renewable energy sources in the world that are to a large extent located in developing and emerging market economies.

Utilizing additional funds for clean energy investment is not only a national effort but also an international effort. It is a test for developing and emerging market countries but also for IFIs to assist those countries make the transition. So far IFIs have been slow to respond.

The Paris Agreement will be tested in 2020 when countries are supposed to contribute new emissions-cutting plans. At this time, it will be easier to judge if the Paris Agreement has a chance to establish itself as a more enduring international framework. That is, will it substantially reduce the risks of dangerous climate change? We will not know whether Paris meets this test for some time. Now the hard work begins within each country and internationally.

[14] Also known as COP 21.

Chapter 1

THE GLOBAL CROSS BORDER INVESTMENT REGIME AND THE LACK OF AN INTERNATIONAL INVESTMENT ORGANIZATION

The absence of a global government makes global warming an awesomely challenging problem to manage. Nations from around the world gathered in Paris for the United Nations Climate Change Conference, with the goal of a universal agreement on reducing greenhouse gas emissions. Very broad participation is needed to fully address the global "tragedy of the commons" that results when countries fail to take into account the negative impact of their carbon emissions on the rest of the world. Furthermore, free riding by non-participants, if sufficiently widespread, can undermine the political will to action of participating countries (Arezki and Obstfeld, 2015). It remains to be seen what action countries will take after the Paris meeting and how the agreement will be implemented.

The lack of a comprehensive foreign investment regime also makes risk management for cross border clean energy investments more difficult. No comparable international organization exists to deal with cross border investments in the same way as the World Trade Organization does for trade. The lack of an international framework for cross-border investment makes political risk mitigation challenging. Cross border investment is less likely to take place unless proper risk mitigation is possible, feasible and fair for all parties involved.

Robert O. Keohane and David G. Victor (2011) have examined the international climate regime. They argue that "[t]he international institutions that regulate issues related to climate change are diverse in membership and content. They have been crafted at different times, and by different groups of countries. They have been crafted in a context of diverse interests, high uncertainty, and shifting linkages. They are not integrated, comprehensive, or arranged in a clear hierarchy. They form a loosely-linked regime complex rather than a single international regime" (Keohane and Victor, 2011, p. 20). Keohane and Victor argue that "The infeasibility of a strong comprehensive regime makes climate change a very difficult international problem to manage" (Keohane and Victor, 2011, p. 20). Furthermore "The failure of efforts to develop a comprehensive, integrated climate regime reflects resistance to costly policies in rich countries, such as the United States, and in developing countries alike" (Keohane and Victor, 2011, p. 8). They conclude that "At the present juncture, however, both political reality and the need for flexibility and diversity suggest that it is preferable to work for a loosely linked but effective regime complex for climate change" (Keohane and Victor, 2011, p. 20). However, it is not clear how a loosely linked regime would operate and how it would differ from the current system that has not yet produced favorable results.

Currently about 3000 bilateral investment treaties are in force. As Salacuse (2010) states, this emerging regime for international investment significantly differs from other international regimes. Three of the most important are that: (1) the regime has largely been constructed bilaterally, rather than multilaterally; (2) it gives broad scope to private and decentralized decision making; and (3) no multilateral international organization supports the investment regime (Salacuse 2010, p. 463).

In fact, if one considers Keohane's definition of cooperation as the "co-ordinated mutual adjustment of states policies yielding benefits to participants"[15] one can also argue that institutionalized cooperation in the field of foreign investment does not exist.

The current investment regime has been founded on the assumption that it will increase international investment, which will then lead to increased prosperity and economic development (Salacuse 2010, p. 468). However, it has been questioned whether investment treaties have in fact increased investment flows to poor countries.

The World Bank, for example, has stated that "[e]ven the relatively strong protections in [bilateral investment treaties] do not seem to have increased flows of investment to signatory developing countries" (World Bank, 2003, p. xvii).

[15] Robert Keohane, The Analysis of International Regimes, in Regime Theory, *supra* note 21, at 23.

Given the fragmented investment regime, it is difficult to identify the specific hegemon that has advanced and maintained the investment treaty system. As Salacuse (2010) has argued "…capital-exporting countries have acted as a collective hegemon to create and maintain the investment regime and thereby maintain their global economic advantage, particularly in relation to developing, capital-importing nations" (Salacuse, 2010, p. 434). Indeed, capital-exporting countries have been the primary force driving negotiation of bilateral investment treaties on which the current investment regime is based. Part of the reason is that after World War II capital-exporting countries felt the need to protect the investments of their nationals. The need for such protection was heightened during the decolonization of territories that had previously been under the control of capital-exporting states.

Why have the nations of the world been willing to negotiate bilateral investment treaties in such large numbers over the last fifty years instead of negotiating global agreements? From a technical perspective it is of course less complicated to negotiate a bilateral agreement than a global treaty that must accommodate the interests of many countries. From a political perspective, given the asymmetric nature of bilateral negotiations between a strong developed country and a usually much weaker developing country, the bilateral setting allows the developed country to use its power more effectively than does a multilateral setting. For example, in multilateral negotiations, developing countries have the opportunity to cooperate with like-minded states to increase their power in negotiations. This would be impossible in bilateral negotiations (Salacuse, 2010, p. 464).

When discussing international investors' efforts to manage political risks, Wells (2005) considers four options: (i) international arbitration, (ii) official political risk insurance, (iii) home government support, and (iv) finally, official credit.

In the absence of a global investment agreement such as the General Agreement on Tariffs and Trade and later the World Trade Organization, investors have turned to piecemeal solutions when protecting their rights in risky countries. According to Wells, "[t]hese agreements set out rules for trade, but they provided few rules for investment.....They did nothing to manage the political risks that could hinder foreign investment. Starting with the aborted International Trade Organization of the immediate post-World War II era, several efforts to create a similar global framework for investment came to naught. The history of failure did not encourage renewed efforts to create a comprehensive approach" (Wells, 2005, p. 89-90). Furthermore, Wells states that "[t]he resulting system, however, was not the product of any grand design but the result of uncoordinated steps by various parties. Certainly, some of the problems of the new system derive from the lack of a single framework; even more important problems can be attributed to the lack of explicit negotiation and mutual acceptance among the affected parties" (Wells, 2005, p. 89-90).

This failure described by Wells is especially serious if one considers clean energy projects that tend to be large, capital intensive and long term. An additional challenge is that energy resources are to a large extent located in developing and emerging countries that are also currently growing faster than high income industrialized countries, both in terms of Gross Domestic Product and population, and thus also energy use. When host governments in developing and emerging countries cannot make credible long term commitments to foreign investors, those investors will tend to avoid these projects. This becomes especially troubling during times when a global need exists for transition to clean energy projects. As Wells points out, "[t]he need to satisfy the demand for security grew as the international community became increasingly eager to encourage private foreign investors to build infrastructure – roads, power plants, water systems – in the developing world" (Wells, 2005, p. 89). "Without external protection, direct investors in these industries would have to be very brave, or perhaps ignorant, to enter these industries, where they would have little bargaining power once their capital was committed" (Wells, 2005, p. 89).

As mentioned above, Wells (2005) noted four options when discussing international investors' efforts to manage political risks: (i) international arbitration, (ii) official political risk insurance, (iii) home government support, and (iv) official credit.

From the point of view of project investors, option (i), international arbitration, faces the objection that engaging in cross border investment and relying on favorable international arbitration in the event of dispute is not a predictable means of mitigating political risks for foreign investors, but it allows them to escape domestic courts. This can be a lengthy and costly process both for the host country and the foreign investor. The World Bank Group-sponsored International Centre for Settlement of Investment Disputes (ICSID)[16] is for example a leading international arbitration institution devoted to international investment dispute settlements. ICSID only facilitates resolution of investor-

[16] The International Centre for Settlement of Investment Disputes (ICSID) is an autonomous international institution established under the Convention on the Settlement of Investment Disputes between states and nationals of other states. It currently has more than 140 member states. The convention sets forth ICSID's mandate, organization and core functions. The intended purpose of ICSID is to provide facilities for conciliation and arbitration of investment disputes. ICSID is one of five institutions that make up the World Bank Group (World Bank, 2015a).

state disputes. In the case of energy projects, the Energy Charter Treaty organization and secretariat whose concern is trade and investment in the energy sector also provides for an investment dispute settlement mechanism.[17] The Energy Charter Treaty applies to all types of energy, including geothermal energy projects. However, the Energy Charter Treaty can only be used in relation to investors of states and host states that have ratified/acceded to it and it does not enjoy global membership. Arbitration can also be based on bilateral investment agreements that include an arbitration clause.

Option (iii), home government support, may be viable for investors from large countries such as the USA or e.g., larger EU member states but faces the objection that home government support does not sound as promising for investors from less powerful countries such as Iceland. In fact, being an investor from a small country only adds to the risks as small countries can only be expected to wield limited leverage in the event of a dispute with a host government in a developing country that could be a much larger country. Options (ii) i.e., official political risk insurance and (iv) i.e., official credit could be a possibility for investors from smaller as well as larger countries to consider when making a foreign investment decision, if feasible venues for cooperation with, e.g.,, IFIs and export credit agencies can be found.

If official political risk insurance is chosen from a multilateral institution such as the World Bank's Multilateral Investment Guarantee Agency (MIGA)[18]

---

[17] The Energy Charter Treaty contains a comprehensive system for settling disputes on matters covered by the Treaty. The two basic forms of binding dispute settlement are state-state arbitration on the interpretation or application of almost all aspects of the Treaty (except for competition and environmental issues), and investor-state arbitration for investment disputes. See http://www.energycharter.org/what-we-do/dispute-settlement/overview/According to an email received from Alejandro Carballo, Legal Counsel, Energy Charter Secretariat, on 17 August 2015 the Energy Charter Treaty applies to all types of energy, including geothermal energy projects. Furthermore, it applies only in relation to investors of states and host states that have ratified/acceded to it. Existing Contracting Parties are: Afghanistan, Albania, Armenia, Australia*, Austria, Azerbaijan, Belarus°, Belgium, Bosnia and Herzegovina, Bulgaria, Croatia, Cyprus, Czech Republic, Denmark, Estonia, the European Union and Euratom, Finland, France, Georgia, Germany, Greece, Hungary, Iceland, Ireland, Italy, Japan, Kazakhstan, Kyrgyzstan, Latvia, Liechtenstein, Lithuania, Luxembourg, Malta, Moldova, Mongolia, The Netherlands, Norway*, Poland, Portugal, Romania, Russian Federation*, Slovakia, Slovenia, Spain, Sweden, Switzerland, Tajikistan, The former Yugoslav Republic of Macedonia, Turkey, Turkmenistan, Ukraine, the United Kingdom, Uzbekistan.

° did not ratify the Energy Charter Treaty, but applies it provisionally.

* did not ratify the Energy Charter Treaty.

[18] The Multilateral Investment Guarantee Agency (MIGA) is one of the five organizations included in the World Bank Group. Established in 1988, its intended mission is to promote foreign direct investment into developing and emerging market countries by providing risk insurance/guarantees and credit enhancement and thereby to help support economic growth, reduce poverty, and improve people's lives. MIGA's operational strategy in intended to

or from a bilateral insurer such as the USA's Overseas Private Investment Corporation,[19] it is important that the insurance policy truly helps the private sector mitigate against risks as this can facilitate capital flows to places where funding is needed. But in addition to the interest of the insured one also needs to consider the interests of the host country and the behavior of the insurer.

From the point of view of the host country as well as the insurer, some issues exist to be concerned about, including two types of moral hazard. First, in the event of a dispute, an insured investor has less incentive to renegotiate a contract than an investor without insurance. The investor may be tempted to walk away from a project without considering adjustment to the terms of the contract in the host country and simply claim the insurance. This can be a serious problem for a host country that is for example faced with a financial crisis such as the regional crisis in Asia in 1997/98 or the global crisis in 2008.

Second, the insurer may have an incentive to pay a claim to a foreign investor since its next move can be to make a claim on the host country and collect its money. The World Bank as insurer is often in a strong position *vis-à-vis* the host country since it may be supporting other projects and programs that will be affected if the host country does not reimburse the bank that just paid out a claim.

But where is the international investment regime heading? One can argue that the world is gradually moving towards an international arrangement on cross border investments. This started with bilateral investment treaties (currently about 3000), then with regional arrangements such as the EU and the North American Free Trade Agreement that have provisions for cross border investments. The next step in this evolution is an inter-regional agreement such as the Trans-Pacific Partnership[20] and the Transatlantic Trade and Investment Partnership[21] currently under discussion and negotiation.

---

attract investors and private insurers into difficult operating environments. MIGA is intended to focus on insuring investments in the areas where it can make the greatest difference: (i) countries eligible for assistance from the International Development Association, (ii) fragile and conflict-affected environments, (iii) complex projects that can be transformational, especially in infrastructure and extractive industries, and (iv) middle-income countries where the agency can have an impact. MIGA offers coverage for five non-commercial/political risks. Coverage can be purchased individually or in combination: (i) currency inconvertibility and transfer restrictions, (ii) expropriation, (iii) war, terrorism and civil disturbance, (iv) breach of contract and (v) nonhonoring of financial obligations (World Bank, 2015a).

[19] For Political Risk Insurance offered by the Overseas Private Investment Corporation, see https://www.opic.gov/what-we-offer/political-risk-insurance.

[20] Trans-Pacific Partnership https://ustr.gov/tpp/.

[21] Transatlantic Trade Investment Partnership http://ec.europa.eu/trade/policy/in-focus/ttip/.

It is possible to view the move from bilateral to regional agreements and then inter-regional agreements as a way to proceed toward a global investment regime like the General Agreement on Tariffs and Trade and then the World Trade Organization on trade. However, one could also view this as four stage evolution: (i) bilateral, (ii) regional, (iii) inter regional, (iv) global as taking the attention away from global efforts. The different language used in different agreements can make consolidation of those agreements into a global arrangement difficult.

What could the reasons be for lack of progress towards a global arrangement? One reason could be that it is easier to negotiate with a smaller number of countries participating in regional agreements such as the North American Free Trade Agreement, the Trans-Pacific Partnership and the Transatlantic Trade and Investment Partnership than in a global setting. Geo-political tensions can play a role such as US exclusion of China from the Trans-Pacific agreement that some view as intentional. The USA may also be seen as exercising power over its smaller neighbors without China being involved.

Multilateral investment agreements can be beneficial to developing countries for several reasons including greater availability of the appeals process (e.g., if one compares some bilateral investment treaties with proposals for the Transatlantic Trade and Investment Partnership) as well as consistency of language. Developing countries could also secure a better deal in a multilateral arrangement that involves more than one rich country. This is because rich capital exporting countries will be protecting themselves against cases involving other rich capital exporting countries. This could imply better agreements for developing countries than a bilateral agreement between a rich and a poor country. In other words, rich countries are not as likely to dominate in regional negotiations that include several rich countries or in a global setting where many rich countries are participating.

In addition to the current trend for more multilateral arrangements (regional and inter-regional) another possibility is that bilateral agreements will improve. Current cases of countries such as Indonesia withdrawing from the bilateral investment treaty system suggest that as a consequence we might achieve better agreements where multilateral arrangements such as the Transatlantic Trade and Investment Partnership might serve as a model for better bilateral investment treaties.

CROSS BORDER CLEAN ENERGY INVESTMENT AND THE CHALLENGING BUSINESS AND INVESTMENT ENVIRONMENT IN DEVELOPING AND EMERGING ECONOMIES

Utilizing clean energy sources in the world is not merely a local or regional issue. It is a global issue affecting both rich and poor countries. Many developing countries and emerging market countries in Africa, Asia and Latin America possess large geothermal energy resources, but most are only in the early stages of development in making use of those resources. Future growth in demand for electricity is also likely to be strongest in these high-growth emerging regions. A transition to clean energy could be of great benefit to those regions as well as having global implications environmentally, and in the battle against climate change. However, it is not sufficient to have the natural resources and potential growth in future demand. Funding is also needed and energy investments tend to be large, capital intensive and with long repayment periods. In addition to government and donor contributions, private sector participation is also important, resulting in public private partnerships.

But this does not mean that private sector funds flow to developing countries, which often have challenging business and investment environments that discourage private sector cross border investment, especially those with long repayment periods, as is the case with geothermal and hydro-power projects.

The World Bank Group has published the so called "Doing Business" report annually since 2002. The report presents quantitative indicators that can be compared across 189 economies—from Afghanistan to Zimbabwe—and over time. In its 2014 report the World Bank ranked 189 countries (World Bank, 2013a). The best performing country is ranked 1 and the worst performing country is ranked 189. The report uses eleven indicators[22] that are then combined into one overall indicator: Ease of doing business.

Table 2 below summarizes some selected indicators for the business and investment climate in some developing and emerging countries that the World

[22] Regulations affecting 11 areas in the life of a business are covered in the 2014 "Doing Business" report: starting a business, dealing with construction permits, getting electricity, registering property, getting credit, protecting investors, paying taxes, trading across borders, enforcing contracts, resolving insolvency, and employing workers. The employing workers data were not included in the ranking on the ease of doing business in the 2014 report.

Bank Group[23] expects will have significant additions in installed geothermal capacity in the future (World Bank, 2012b, p. 31). As Table 2 shows, most of these countries face a challenging business and investment environment. African countries tend to be low income whereas Asian and Latin American countries tend to be middle income. In most countries companies have difficulty obtaining electricity and they typically score low on indicators that are important for cross border energy investment, such as dealing with construction permits, protecting investors, enforcing contracts and trading across borders.

Investors, especially those considering long term investments, will avoid difficult business and investment environments unless risks can be managed. Proper risk mitigation for operators and investors is thus a major challenge in most of the African countries listed below as well in many developing and emerging countries in Asia and Latin America.

It must be noted that the data in Table 2 below only provide some indication for investors. The "Doing Business" report does not measure the full range of factors, policies and institutions that affect the quality of the business environment in an economy or its national competitiveness. It does not, for example, capture aspects of security, the prevalence of bribery and corruption, market size and macroeconomic stability.[24] A more thorough review would be needed for a company to make a decision about cross border engagement. Nevertheless, the "Doing Business" report is a useful start to begin identifying challenges in each country. Gaining a fuller understanding of the broader business environment and a broader perspective on policy challenges requires combining insights from "Doing Business" with data from other sources, such as World Bank Enterprise Surveys.[25] Other indicators more specific to the regulatory environment for energy investment also need to be studied and assessed. Moreover, political or non-commercial risks are associated with cross border investments and those risks much be managed. These will be discussed in the following chapter on international financial institutions.

[23] The World Bank Group represents five institutions. These are: (i) the International Bank for Reconstruction and Development (IBRD), established in 1944, (ii) the International Development Association (IDA), established in 1960, (iii) the International Finance Corporation (IFC), established in 1956, (iv) the Multilateral Investment Guarantee Agency (MIGA), established in 1988, (v) the International Centre for Settlement of Investment Disputes (ICSID), established in 1966. Four of these institutions issue insurance or guarantees, i.e.: IBRD, IDA, IFC and MIGA.

[24] Including whether the government manages its public finances in a sustainable way.

[25] See, http://www.enterprisesurveys.org.

Table 2. According to the World Bank Group, significant additions in installed geothermal capacity can be expected in the following African, Asian Pacific and Latin American countries

Region	Income Level	Ease of Doing Business Rank	*Getting Electricity*	Dealing with Construction Permits	Protecting Investors	Enforcing Contracts	Trading Across Borders
Africa							
Tanzania	Low income	145	*102*	177	98	42	139
Eritrea	Low income	184	*95*	189	115	67	170
Sudan	Lower middle income	149	*113*	167	157	154	155
Somalia	N/A	N/A	*N/A*	N/A	N/A	N/A	N/A
Malawi	Low income	171	*183*	173	80	145	176
Zambia	Low middle income	83	*152*	57	80	120	163
Burundi	Low income	140	*161*	126	34	177	175
Rwanda	Low income	32	*53*	85	22	40	162
Uganda	Low income	132	*178*	143	115	117	164
Congo, Dem. Rep.	Low income	183	*142*	90	147	177	171
Mozambique	Low income	139	*171*	77	52	145	131
Madagascar	Low income	148	*187*	157	68	160	115
Comoros	Low income	158	*109*	44	138	159	146
Mauritius	Upper middle income	20	*48*	123	12	54	12
Asia Pacific							
Malaysia	Upper middle income	6	*21*	43	4	30	5
Papua New Guinea	Lower middle income	113	*24*	165	68	168	134
Latin America							
Guatemala	Lower middle income	79	*34*	61	157	97	116
Honduras	Lower middle income	127	*125*	83	170	182	84
Panama	Upper middle income	55	*16*	62	80	127	11
Colombia	Upper middle income	43	*101*	24	6	155	94
Ecuador	Upper middle income	135	*138*	64	138	99	122
Bolivia	Lower middle income	162	*128*	136	138	131	126

Source: World Bank, 2012b and 2013a

Chapter 2

THE WORLD BANK GROUP: LOANS VS. GUARANTEES?

The World Bank Group now represents five distinct institutions. These are: (i) the International Bank for Reconstruction and Development, established in 1944, (ii) the International Development Association, established in 1960, (iii) the International Finance Corporation, established in 1956, (iv) the Multilateral Investment Guarantee Agency, established in 1988, (v) the International Centre for Settlement of Investment Disputes, established in 1966. For more detail about each institution see box 1 below.

Box 1. The five institutions of the World Bank Group.[26]

The International Bank for Reconstruction and Development (IBRD).

Established in 1944, the IBRD aims to reduce poverty in middle-income countries and creditworthy poorer countries by promoting sustainable development through loans and guarantees, risk management products, and analytical and advisory services. The IBRD is structured like a cooperative that is owned and operated for the benefit of its 188 member countries. The IBRD raises most of its funds on the world's financial markets and has become one of the most established borrowers since issuing bonds in 1947. The IBRD's financial strength has enabled it to borrow at relatively low cost.

[26] The World Bank Group consists of IBRD, IDA, IFC, ICSID and MIGA. The World Bank usually only includes IBRD and IDA.

(Continued)

The International Development Association (IDA).

Established in 1960, the IDA aims to reduce poverty in the poorest countries by providing loans and guarantees for programs that boost economic growth, reduce inequalities, and improve people's living conditions. The IDA lends money on concessional terms. This means that the IDA charges low to no interest and repayments take place over periods of 25 to 40 years, including five to ten year grace periods. The IDA provides guarantees to help attract private financing.

The International Finance Corporation (IFC).

Established in 1956, the IFC is the largest global development institution focused exclusively on the private sector in developing and emerging countries. The IFC is owned by 184 member countries, a group that collectively determines its policies. With work in more than 100 developing and emerging countries, the IFC aims to help companies and financial institutions create jobs, generate tax revenues, improve corporate governance, and contribute to their local communities. According to the World Bank Group the IFC has five stated priorities: (i) strengthening the focus on frontier markets; (ii) addressing climate change and ensuring environmental and social sustainability, (iii) addressing constraints to private sector growth in infrastructure, health, education, and the food supply chain, (iv) developing financial markets, and (v) building long-term client relationships in emerging markets.

The Multilateral Investment Guarantee Agency (MIGA).

Established in 1988, MIGA's mission is to promote foreign direct investment in developing and emerging market countries by providing risk insurance/guarantees and credit enhancement, thereby helping to support economic growth, reduce poverty, and improve people's lives. MIGA's operational strategy is intended to attract investors and private insurers into difficult operating environments. MIGA is intended to focus on insuring investments in the areas where it can make the greatest difference: (i) countries eligible for assistance from the International Development Association (IDA), (ii) fragile and conflict-affected environments, (iii) complex projects that can be transformational, especially in infrastructure and extractive industries, and (iv) middle-income countries where the agency can have an impact. MIGA offers coverage for five noncommercial/political risks. Coverage can be purchased individually or in combination: (i) currency inconvertibility and transfer restrictions, (ii) expropriation, (iii) war, terrorism and civil disturbance, (iv) breach of contract and (v) nonhonoring of financial obligations.

The International Centre for Settlement of Investment Disputes (ICSID).

Established in 1966, ICSID is an autonomous international institution established under the Convention on the Settlement of Investment Disputes between states and nationals of other states. Currently more than 140 states are members. The Convention sets forth ICSID's mandate, organization and core functions. The intended purpose of ICSID is to provide facilities for conciliation and arbitration of investment disputes. ICSID is one of five institutions that make up the World Bank Group.

Source: World Bank, 2015a.

All five World Bank Group institutions can play an important role in supporting clean energy investments including geothermal projects. They can assist with funding (e.g., lending and equity instruments, and grants), risk mitigation (e.g., guarantees and insurances) as well as with dispute settlement. Furthermore, they can advise governments on policy and support projects with technical assistance.

Table 3 below summarizes World Bank Group investments in geothermal projects over about three decades. As can be seen from the table, the World Bank has engaged in the African region (Djibouti, Ethiopia and Kenya), in Asia (Indonesia and the Philippines), and finally in Europe (Armenia, Lithuania and Poland). This must be considered a modest engagement, especially if one considers the investment made, totaling US$1.2 billion. This amounts at best to a few cents per capita per year in countries with geothermal potential.

The main World Bank Group institutions involved are the International Development Association, which provides concessional lending to governments of low income countries, and the International Bank for Reconstruction and Development, which provides lending to middle income countries. The Global Environmental Facility[27] has also made contributions. The Global Environmental Facility is a partnership for international cooperation in which 183 countries cooperate with international institutions, civil society organizations and the private sector to address global environmental issues.

[27] Established in 1991, the Global Environmental Facility is a partnership for international cooperation in which 183 countries cooperate with international institutions, civil society organizations and the private sector to address global environmental issues (World Bank, 2015a).

Table 3. Geothermal investments of the World Bank over about three decades (US$ million)

Date	Country	Region	Project ID	Project Name	Total Amount of Project US$	World Bank Contribution US$	Source
2013	Djibouti	Africa	P127143, P127144	Geothermal Power Generation Program	31.23	13.14	IDA (US$6 million), Global Environmental Facility (US$6.04 million), Energy Sector Management Assistance Program (US$1.1 million)
1984	Djibouti	Africa	P000609	Djibouti - Geothermal Exploration Project	17.4	6	IDA
1989	Djibouti	Africa	P000612	Geothermal Development Project	38.04	9.2	IDA
2014 (under preparation)	Ethiopia	Africa		Geothermal Sector Development Project	236	150 (TBC)	IDA (plus US$24.5 million from Scaling-up Renewable Energy in Low Income Countries Program)
2010	Ethiopia	Africa	P120172	Additional Financing for Energy Access Project	30	15	IDA
1983	Ethiopia	Africa	P000696	Ethiopia - Petroleum Exploration Promotion and Geothermal Reconnaissance Project	9.84	7	IDA

Date	Country	Region	Project ID	Project Name	Total Amount of Project US$	World Bank Contribution US$	Source
2014 (under preparation)	Kenya	Africa	P129910	Menengai Geothermal	848	115	IDA (plus US$15 million from Scaling-up Renewable Energy in Low Income Countries Program)
2012	Kenya	Africa	P122671	Private Sector Power Generation Support Project	212	31	IDA – Partial Risk Guarantee
2010	Kenya	Africa	P103037	Electricity Expansion Project	1035	120	IDA
2004	Kenya	Africa	P083131	Energy Sector Recovery Project	55	25.5	IDA
1988	Kenya	Africa	P001314	Kenya Geothermal Development and Energy Pre-investment Project	66.95	40.7	IDA
1984	Kenya	Africa	P001290	Kenya - Geothermal Exploration Project	24.5	24.5	IDA
1983	Kenya	Africa	P001287	Olkaria Geothermal Power Expansion Project	41.6	12	IBRD
1980	Kenya	Africa	P001275	Olkaria Geothermal Power Project	40	40	IBRD

Table 3. (Continued)

Date	Country	Region	Project ID	Project Name	Total Amount of Project US$	World Bank Contribution US$	Source
1978	Kenya	Africa		Olkaria Geothermal Engineering Project	15.5	9	IBRD
2011	Indonesia	East Asia and the Pacific Region	P113078	Geothermal Clean Energy Investment Project	574	175	IBRD (plus US$125 million from Clean Technology Fund for WB project, US$300 million total Clean Technology Fund)
2011	Indonesia	East Asia and the Pacific Region	P112363	Geothermal Power Support Program		0.9	IBRD
2008	Indonesia	East Asia and the Pacific Region	P099757	Geothermal Power Generation Development Project	9	4	Global Environmental Facility
1994	Philippines	East Asia and the Pacific Region	P004607, P004404	Leyte Luzon Geothermal Project	1333.6	257	IBRD (US$227 million), Global Environmental Facility (US$30 million)
1994	Philippines	East Asia and the Pacific Region	P004584	Leyte Cebu Geothermal Project	434.8	211	IBRD

Date	Country	Region	Project ID	Project Name	Total Amount of Project US$	World Bank Contribution US$	Source
1988	Philippines	East Asia and the Pacific Region	P004555	Bacon Manito Geothermal Development Project	289.5	96.1	IDA
1982	Philippines	East Asia and the Pacific Region	P004493	Geothermal Exploration Project	58.75	8.23	IDA
2009	Armenia	Europe and Central Asia Region	P112408	Armenia Geothermal Project (Geofund)		1.5	Global Environmental Facility
1996	Lithuania	Europe and Central Asia Region	P036011, P035758	Lithuania Klaipeda Geothermal	18	12.8	IBRD (US$5.9 million), Global Environmental Facility (US$6.9 million)
2000	Poland	Europe and Central Asia Region	P037339	Podhale Geothermal District Heating & Environment Project	99.3	38.2	IBRD
2000	Poland	Europe and Central Asia Region	P057993	Podhale Geothermal District Heating & Environment Project		5.4	Global Environmental Facility
				Total	**5.518**	**1.237**	

Source: World Bank 2014 email dated 17 February 2014.

Guarantees and insurances to support private sector participation and capital mobilization are not listed in the above table[28] although four of the five World Bank Group institutions offer such instruments. Those institutions are: the International Bank for Reconstruction and Development, the International Development Association, the International Finance Corporation and the Multilateral Investment Guarantee Agency.

When the World Bank was established at Bretton Woods, views naturally differed about its role and the instruments it should use. The world was different from what it is today with one country, the United States, dominating and acting as a hegemon while Europe and Japan were in desperate need of reconstruction. This led to the establishment of the International Bank for Reconstruction and Development (IBRD). The initial intention was that the IBRD would focus on post World War II reconstruction but it then turned its attention to economic development when colonies gained independence. Today the hegemonic role of the USA is less clear and its main rival, China, is leading efforts to establish two new banks, the Asian Infrastructure Investment Bank and the New Development Bank, as will be discussed later in this book.

Galvin and Rodrik (1995) discuss the debate on the World Bank at Bretton Woods in their article entitled: *The World Bank in a Historical Perspective.* When discussing the views of the Bretton Woods delegates regarding the bank's then proposed guarantee functions they state that "A policy decision that decisively shaped the Bank's structure in this regard was whether the Bank should primarily operate by participating in and guaranteeing private loans or should instead make loans on its own account, from its capital or borrowed resources. This issue was discussed at much length in the negotiations that preceded the Bretton Woods agreements. The Europeans, and Keynes in particular, argued strongly that the guarantee function should predominate" (Gavin and Rodrik, 1995, p. 330). According to this the most distinguished figure at the Bretton Woods conference, John Maynard Keynes, felt that guarantees should be the key instrument, or at least among the key instruments in the reconstruction of post-World War II Europe. When discussing the views of the Bretton Woods delegates, Galvin and Rodrik (1995) further state "Though they obviously felt the need for public presence in development lending, the Bank's creators were also keenly aware that, whatever its deficiencies, private investment would have to remain an essential part of the international lending scene. And policymakers were anxious to ensure that the Bank proved

[28] Except for a Partial Risk Guarantee from the International Development Association in Kenya in 2012.

complementary to private international lending, though it was not obvious how to ensure this, or how best to structure the Bank's relationship to the private lending community" (Gavin and Rodrik, 1995, p. 331). But this was not to be. Although "[t]here was little disagreement about the desirability of the Bank using its power to provide full or partial guarantees of private loans, and there is considerable evidence to suggest that these were expected to be important [...], within a year of the Bank's inauguration, it was obvious that the Bank would place almost total emphasis on direct lending rather than guarantees or participation in private loans" (Gavin and Rodrik, 1995, p. 330) and as Krueger (1997) pointed out that "In practice, the IBRD has lent and made little use of its guarantee powers" (Krueger, 1997, p. 8).

To date this continues to be the situation. Although the World Bank's mandate to provide guarantees is rooted in its Articles of Agreement, guarantees continue to remain an instrument of minor importance within the World Bank Group. Nevertheless, discussion about the potential for their growing importance has taken off from time to time within the World Bank. A notable example of this discussion is a memorandum signed by the World Bank president, Lewis T. Preston, entitled "Mainstreaming of Guarantees as an Operational Tool of the World Bank", sent to the World Bank´s Executive Directors on July 14, 1994 (World Bank, 1994).

This memorandum recognized guarantees as an important complement to World Bank loans and considered them a highly flexible instrument that could be customized to suit varying country and project circumstances. In fact, as the memorandum stated, the World Bank's Articles of Agreement had envisaged the International Bank for Reconstruction and Development to be primarily in the guarantee business (World Bank, 1994) although guarantees never found a widespread role in World Bank operations.

Written in 1994, the memorandum recognized that recent developments in financing requirements of many of the World Bank borrowing countries indicated that guarantees should become a much more significant form of World Bank support. Among the reasons mentioned was that the financing needs for development of infrastructure in developing countries were estimated to be very large and well beyond the capacity of official sources alone to support. Besides, many World Bank borrowers were turning to the private sector to invest in infrastructure projects through a variety of private ownership arrangements which would attract not only private sector resources but also allow a greater transfer of technical and operational assistance (World Bank, 1994).

According to the memorandum, the objective of the World Bank in offering guarantees was to help mobilize private sector financing for individual projects through targeted and limited support, and to leverage the World Bank‘s participation. “Guarantees are a particularly well-suited form of Bank support for economically important private sector project finance transactions, where they can be targeted to mitigate against specific risks—generally risks of political, regulatory and governmental performance—which in particular projects the private sector may not be in position to absorb or manage” (World Bank, 1994, p. 3).

The comparative advantage of guarantees was seen as most apparent in infrastructure investments that require large amounts of funds with extended maturities to match the long pay-back periods normally associated with such projects (World Bank, 1994). Furthermore, the president’s memo noted that “Even though there have been several successful private financings of infrastructure projects in developing countries in recent years, these have only amounted to a small fraction of the total needs” (World Bank, 1994, p. 3). The same situation applies today, more than 20 years later.

The president’s memo also considered the potential demand for guarantees and reported the results of discussions with several international investors and commercial investment banks. The memorandum states that “Their reactions have been almost uniformly positive towards the prospect of Bank guarantees, particularly to cover government performance risks in private sector infrastructure projects” (World Bank, 1994, p. 7). It was argued that many projects “would simply not materialize within a reasonable time period without Bank support” (World Bank, 1994, p. 7). The view was also expressed that “once the existence of the Bank’s guarantee program became better known and understood by the investors and the financiers, investors would show renewed interest in many more projects in a much wider range of countries than in the past” (World Bank, 1994, p. 7). According to the president’s memorandum, coordination within the World Bank Group “would be strengthened through more structured consultations between each country department and IFC’s infrastructure department and MIGA” (World Bank, 1994, p. 12). Lack of coordination among World Bank Group institutions was thus an issue in 1994 as it is today and will be discussed further in this book.

In terms of discussing the potential to leverage more funds through guarantee instruments instead of loans and equity contributions the memo stated as follows: “In practice, since partial risk guarantees represent a contingent obligation to disburse, the likelihood that the [World] Bank would disburse against a guarantee would depend on the probability of occurrence of the event triggering the call, which is likely to be less than one. It should, therefore, in principle be possible to support a larger volume of partial risk guarantees than loans within the same country risk exposure" (World Bank, 1994, p. 16). This must be classified as a remarkably conservative statement by the World Bank president and confirms the potential to leverage larger funds with guarantee instruments than via loans and equity contributions only.

The memorandum concluded as follows: "Guarantees potentially represent a useful and flexible instrument of Bank support for specific projects and country assistance strategies. Recent developments in the financing needs of Bank borrowers and their increasing interest in promoting private sector investments in infrastructure open the possibilities of utilizing Bank guarantees as a mainstream instrument, complementing the support provided by MIGA and IFC" (World Bank, 1994, p. 18).

The World Bank started issuing guarantees only in 1983 for the purpose of attracting private co-financing for World Bank-supported projects. In 1994, the World Bank board approved specific policy provisions for the use of partial risk guarantees and partial credit guarantees for private and public sector projects in IBRD-eligible countries.[29] Partial risk guarantees became available for projects in IDA-only countries in 1997 in the form of IBRD guarantees for enclave projects[30], and IDA partial risk guarantees for private sector projects.[31] Policy-based guarantees were introduced in 1999 for well-performing IBRD borrowers.[32] In 2002, management summarized the World Bank guarantee policy provisions approved by the executive directors since 1994 in an operational policy statement and issued an accompanying bank procedures statement (World Bank, 2013c, p. 4). The effectiveness of World Bank current risk mitigation instruments will be discussed later in this book.

---

[29] See *Mainstreaming of Guarantees as an Operational Tool of the World Bank.* Washington DC, R94-145.

[30] See *Use of IBRD Guarantees to Support Private Enclave Project in IDA-Only Countries,* (R97-85, IDA/R97-36), 28 April 1997.

[31] See *A Proposal for IDA Guarantees in IDA-Only Countries* (IDA/R97-135), November 20, 1997.

[32] See *World Bank Policy-Based Guarantees* (R99-53), 20 April 1999.

Chapter 3

INTERNATIONAL FINANCIAL INSTITUTIONS AND RISK MITIGATION

When discussing the role of IFIs, the Stern Review on the economics of climate change states that: "The International Financial Institutions have an important role to play in accelerating this process: the establishment of a Clean Energy Investment Framework by the World Bank and other multilateral development banks offers significant potential for catalysing and scaling up investment flows" (Stern Review, 2006, p. xxiv). But what are those multilateral development banks?

In this book multilateral development banks are called international financial institutions (IFIs). IFIs include the main global development institution, the World Bank,[33] whose foundations were laid at the Bretton Woods Conference in New Hampshire in the USA during World War II. IFIs also include banks with a regional focus such as the African Development Bank (established in 1964), the Asian Development Bank (established in 1966), the European Bank for Reconstruction and Development (established in 1991), and the Inter-American Development Bank (established in 1959). The European Union's European Investment Bank (established in 1958) also plays an important role, as does the Nordic Investment Bank (established in 1975). Those institutions deploy various instruments including direct lending as well as guarantee instruments that can be used for capital mobilization in projects involving the private sector.

As to engagement in geothermal development, the World Bank Group is by far the most active institution, see Figure 1 below.

[33] Then called the International Bank for Reconstruction and Development (IBRD).

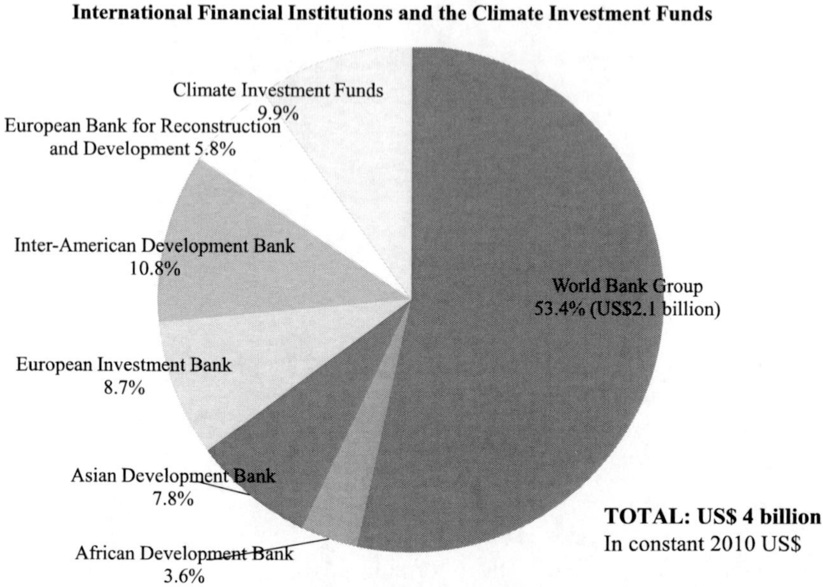

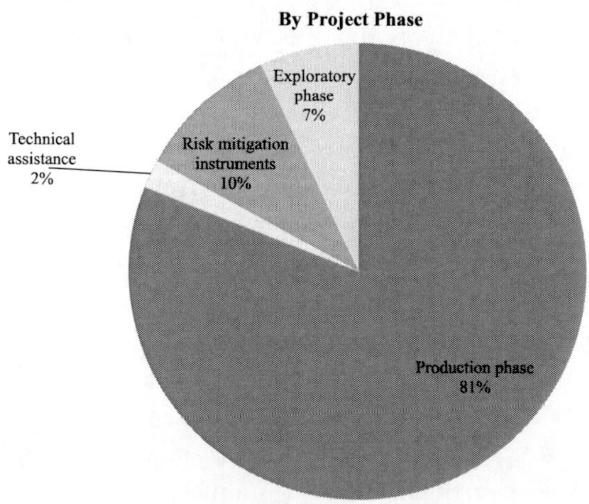

Figure 1a, based on recent World Bank Group data, shows cumulative IFI and Climate Investment Funds lending for geothermal development from 1977 to the present. Figure 1b also shows growing, but still insufficient, IFI/donor support directed to the exploration phase discussed later in this book. Risk mitigation instruments are also beginning to appear for capital mobilization.

Source: World Bank, 2015b.

Figure 1a and 1b. Cumulative IFI and Climate Investment Funds Lending for Geothermal Development (1977 –present).

Clean Energy Investment and Political Risk Mitigation

Concerns about investment environments and perceptions of political risk often inhibit cross border foreign investment, resulting in most investment flows going to a few countries, leaving the world's poorest economies largely ignored. This is especially true during times of economic and financial crisis. The limited number of investors engaging in risky environments might also be tempted to invest only when quick payback periods are possible. IFIs can have an important role to play here and a responsibility to offer effective venues and viable risk mitigation instruments, including guarantees and insurance instruments. This is especially true of large and long-term investments such as energy infrastructure, including geothermal and hydropower plants.

For large and long-term infrastructure projects investors must pay considerably more attention to political risk management issues than for small and short-term engagements. Risk mitigation can reduce the cost of funding projects and facilitate longer loan periods. Risk varies inversely with knowledge and investors can thus also reduce their risk by researching the investment environment they are entering, in this way increasing their level of knowledge about the environment they are entering and engaging in. However, political circumstances can change in unexpected ways. Political risk insurance, especially from IFIs, can be effective and act as an effective deterrent against host government interference with insured private investments.

IFIs offer a number of financial and risk management instruments that can be useful for public private partnerships (PPPs).[34] These include loans, equity investments and guarantees or insurance against political risk (non-commercial risk). As stated above, among the IFIs active in this area are: (i) the World Bank Group, (ii) the European Bank for Reconstruction and Development, (iii) the Asian Development Bank, (iv) the Inter-American Development Bank, (v) the African Development Bank, (vi) the European Investment Bank, and (vii) the Nordic Investment Bank. These institutions are well placed to mitigate risks at competitive prices. As Wells states, "official insurance can be priced low, since

[34] In addition to these services, IFIs often engage in policy dialogue with governments of developing and emerging market economies to improve economic policy and management. This includes reforms to improve the business and investment climate for the private sector, to promote business activities, and to encourage foreign direct investment. IFIs also provide loans and credit to government-led projects in developing countries and emerging markets that are subject to international competitive bidding. This allows private sector firms to participate in bidding and potentially to benefit from public sector projects supported by IFIs.

the threat of sanctions by the organizations backing the insurance sharply reduce[s] the chances of the events being insured against occurring" (Wells, 2005, p. 91). In terms of trade finance, the role of export credit agencies can also be important, as will be discussed later in this book.[35]

Key risk issues can be categorized as: political, breach of contract by a government entity, market risk, and default risk.[36] Risk mitigation products can attract new financing resources, reduce costs of capital, and extend maturities by providing coverage for risks that the market is unable or unwilling to bear (Delmon, 2009). These products can attract more private capital to invest in infrastructure. Examples of guarantee products provided by the World Bank Group are partial risk guarantees from the International Bank for Reconstruction and Development (IBRD) or the International Development Association (IDA) and IBRD partial credit guarantees, International Finance Corporation (IFC) partial credit guarantees and political risk insurance from the Multilateral Investment Guarantee Agency (MIGA) (Delmon, 2009). Those risk mitigation instruments allow investors as well as lenders to be compensated in case of certain adverse events, thus reducing the risk and hence also project costs.

In the case of clean energy infrastructure projects, World Bank guarantee products such as partial risk guarantees can be key to success. According to the World Bank, partial risk guarantees "cover commercial lenders for a private sector project against default arising from a government-owned entity failing to perform its obligations. [Partial risk guarantees] can cover changes in [the] law, failure to meet contractual payment obligations, expropriation and nationalization, currency transfer and convertibility, nonpayment of a termination amount, failure to issue licenses in a timely manner, other risks to the extent that they are covered by contractual obligations of a government entity, and noncompliance with an agreed dispute resolution clause. [Partial risk guarantees] can be provided in both IBRD and IDA countries and require a government counter-guarantee" (World Bank, 2009, p. 10).

Regarding the IBRD/IDA partial risk guarantees, the investor receives comfort, improved credit terms, and is not liable for loan repayment. Among the strengths of this instrument is increased government commitment to success of projects, accompanied by the benefits of an ongoing policy dialogue between the World Bank and the host government. Among the weaknesses are sovereign guarantees required in all cases, cumbersome processing and high transaction

[35] See for example the case of Nam Theun 2 in Lao.

[36] For an excellent overview of World Bank risk mitigation products, see also Jeffrey Delmon Chapter 7 (Delmon, 2009).

costs. The demand for this instrument is mainly limited to PPPs and sectors with heavy government engagement (World Bank, 2009, p. 74).

The five institutions of the World Bank Group also include the Multilateral Investment Guarantee Agency (MIGA). MIGA provides guarantees against political risks, i.e., non-commercial risks for investments in emerging markets. It also provides technical assistance and a dispute mediation service. Developing countries would hesitate to take measures that would negatively affect projects that MIGA is involved with because of the concern that it could adversely affect their relationship with the IDA and/or IBRD and possible credit or a loan (see for example West, 1999, p. 29-30).[37]

According to the World Bank, MIGA "offers [political risk insurance] coverage to foreign direct investors for any combination of the following political risks: transfer restriction, expropriation, war and civil disturbance, and breach of contract. MIGA can insure direct equity, quasi-equity, nonequity direct, and other investments. To insure debt, however, it must have an equity link. MIGA guarantees cover new foreign-currency-denominated investments, including "new" investments to existing investments, investments by private for-profit and nonprofit organizations, and publicly owned investors and organizations that operate on a commercial basis. MIGA can cover any freely usable currency, which may include local currency investments/loans. Under certain circumstances, MIGA can cover investments by local investors" (World Bank, 2009, p. 10).

Regarding MIGA political risk insurance, the investor receives comfort, improved credit terms, mediation services and compensation in the event of loss. Among the strengths are flexible coverage of all political risk insurance risks; main product for equity investments; dispute resolution; minimal time and processing. Weaknesses include no comprehensive coverage (commercial risk and political risk cover) and a lengthy process to change Convention limitations (World Bank, 2009, p. 74).

The Asian Development Bank and the other regional development banks also offer risk mitigation instruments that are important for private investors in

[37] The fact that the World Bank Group represents five different institutions that often have multiple projects and programs in recipient countries undoubtedly gives it leverage vis-à-vis governments and the private sector that other institutions do not have to the same degree. One can, however, question if use of this leverage is just and fair in all circumstances. Galvin and Rodrik (1995) comment as follows: "And there is something more than a little schizophrenic about an agency that preempts potential private lenders because they are allegedly too risk-averse, then demands that its loans be senior to any others, thereby shifting most of the risk onto private investors in unrelated projects, whose loans have thereby been subordinated" (Gavin and Rodrik, 1995, p. 331).

emerging markets although they are not discussed in any detail in this chapter. Asian Development Bank risk mitigation instruments can, for example, cover breach of contract. For a power project a breach of this kind may result from failure by a government-owned entity to make payments in accordance with the power purchase agreement between the independent power producer and the user or distributor (Asian Development Bank, 2000, p. 2). Such insurance can be critical for the success of an energy infrastructure project. According to the Asian Development Bank the majority of the partial risk guarantees that it has provided have been private sector-oriented, including partial risk guarantees for PPPs (Asian Development Bank, 2006, p. 5).

IFIs as well as bilateral financial organizations can be a catalyst to support for PPP projects, including in the energy sector. A book published a few years ago by the World Bank, entitled "Public-Private Partnerships in Europe and Central Asia – Designing Crisis-Resilient Strategies and Bankable Projects", comments very cautiously that "working with these institutions may also lengthen the project development process, given specific requirements in terms of environmental and social safeguards requirements and stringent procurement procedures" (Cuttaree and Mandri-Perrott, 2011, p. 59). Another recent book also published by the World Bank entitled: "Doing a dam better: the Lao People's Democratic Republic and the story of Nam Theun 2 (NT2)", is more critical when discussing World Bank cooperation with the private sector. The authors simply state that: "The bad news is that the World Bank is seen as a high-cost/high-hassle partner of last resort. There is therefore a critical need to reduce the costs the private sector incurs for doing business with the World Bank. Doing so will require the World Bank to better understand the constraints under which the private sector works" (Porter and Shivakumar, 2010, p. 22). These comments are especially notable given that the authors have both served as World Bank country directors.

From the above, it seems that a key IFI like the World Bank has some way to go to find an optimal trade-off between excessive safeguards and efficient use of its financial instruments for capital mobilization. This is especially unfortunate given the urgency to construct more clean energy projects in developing and emerging countries.

Chapter 4

HOW INTERNATIONAL FINANCIAL INSTITUTIONS CAN SUPPORT CROSS BORDER CLEAN ENERGY PROJECTS

When private companies engage in cross border investment in developing and emerging markets, capital mobilization and risk mitigation is likely to be among the key challenges they face. Among the most obvious partners to help solve that problem are international financial institutions (IFIs). These often maintain a strong presence in developing and emerging countries and can offer financial instruments such as equity, loans, or guarantee/insurance instruments to support investment projects, as well as technical assistance. The involvement of IFIs could also facilitate participation by private international investment banks, export credit agencies, as well as potential private co-sponsors providing equity capital.

Two World Bank Group institutions, the International Bank for Reconstruction and Development (IBRD) and the International Development Association (IDA), work with host governments requiring host government guarantees. Two other World Bank Group institutions, the International Finance Corporation (IFC) and the Multilateral Investment Guarantee Agency (MIGA), support private sector investment without host government guarantees. IFIs with regional focus such as the African Development Bank, the Asian Development Bank, the European Bank for Reconstruction and Development, the Inter-American Development Bank, and the European Investment Bank offer similar services in the regions within which they operate. Table 4 below summarizes the main instruments offered by various IFIs.

Table 4. International financial institutions, their financial and risk mitigation instruments and technical assistance

Institution	Main Products
Asian Development Bank	Loans, equity, guarantees, advisory services, and syndications.
African Development Bank	Loans, equity, commercial and political risk guarantees, syndications, and technical assistance.
European Bank for Reconstruction and Development	Loans, equity, guarantees, securitized finance, advisory services, and syndications.
European Investment Bank	Loans, equity, guarantees, and technical assistance.
Inter-American Development Bank. Non- sovereign Guaranteed Operations.	Loans, guarantees, grants, technical assistance, and syndications.
International Finance Corporation – World Bank Group	Loans, equity, guarantees, securitized finance, advisory services, and syndications.
Multilateral Investment Guarantee Agency – World Bank Group	Political risk insurance guarantees.

As the table shows, IFIs offer a variety of instruments, including loans and equity as well as guarantees and political risk insurance instruments. Those instruments can help mobilize funding from other sources e.g., loans from private sector commercial banks as well as equity participation from private sector participants

Source: International Finance Corporation, 2011

Capital shortages for cross border energy investments in developing and emerging market countries remain a global problem. It is widely known that investment needs in clean energy in emerging markets and developing countries are huge. In 2009 the International Finance Corporation, for example, estimated that electricity sector investment needs in developing countries from 2007 to 2030 would be US$7.9 trillion (IFC, 2009). That was about half of the Gross National Income of the USA in 2009 when this estimate was made (World Bank, 2010). Interest is growing in public private partnerships (PPPs) among Asian countries, including a notably prominent market in China. The Association of Southeast Asian Nations (ASEAN) has, for example, recently stated that "PPPs are seen to be beneficial in meeting ASEAN infrastructure needs, estimated by the Asian Development Bank at USD 60 billion per annum." (ASEAN, 2014). Meeting those growing needs will be a major challenge.

IFIs contribute billions of dollars every year to support private sector investment in developing and emerging countries. Part of those funds goes to clean energy investments. However, even if all the IFI funds were used for clean energy investment this would only be sufficient to fund a small fraction of the global investment needs for clean energy investment. This is why pooling funds from public and private sources as well as donors and IFIs is necessary.

The international community increasingly emphasizes clean energy investment for environmental reasons and as part of the battle against climate change. To promote those investments IFIs are among the instruments of the international community. Table 5 shows a clear focus on clean and renewable energy, and climate action among most IFIs. Some institutions (the African Development Bank, and International Finance Corporation (IFC) and the Multilateral Investment Guarantee Agency (MIGA) from the World Bank Group) do not mention this specifically, but presumably these kinds of investment would fall under infrastructure investment that they, as well as all the other IFIs, mention as key private sector focus areas.

Partnership with private investors has for a long time formed a central part of IFI support to the private sector. Most IFIs limit their participation in project investment to well under 50 percent, thus requiring partnership with other investors. The structure of IFI finance substantially leverages the capital provided by governments. Not only do IFIs borrow significantly from outside to support their operations, but they also invest in projects alongside private financiers and sponsors. Indicatively, the net result is that one US dollar of capital supplied by governments to an IFI can lead to US$12 of private sector project investment (IFC, 2011). IFIs can thus be key players in supporting capital mobilization in developing and emerging markets.

The Nam Theun 2 hydropower project in Lao PDR, discussed later in this book, is an excellent example of a successful leveraging of multilateral guarantee mechanisms in a difficult business and investment environment, involving the participation of many private investment banks. The risk mitigation instruments used by the World Bank Group were IDA partial risk guarantees and MIGA political risk insurance. The Asian Development Bank also provided a guarantee.[38] The case of the Bujagali hydropower project in Uganda is another notable example, also discussed later in this book.[39]

---

[38] For more detail see references, World Bank, 2005.

[39] For more detail see references, World Bank, 2007.

Table 5. International financial institutions and key private sector focus areas

Institution	Key Private Sector Focus Areas
Asian Development Bank	*Infrastructure*, capital markets, and financial sectors, **with increasing focus on clean and renewable energy**, frontier markets, and underserved economies.
African Development Bank	*Infrastructure*, financial sector, industry, agribusiness, services, regional integration, and inclusive growth.
European Bank for Reconstruction and Development	Industry, commerce and agribusiness, natural resources, **renewable energy**, *infrastructure*, financial institutions, and small and medium-sized enterprises.
European Investment Bank	*Infrastructure*, **energy, climate action,** financial markets, small and medium-sized enterprises, microfinance, and industry.
Inter-American Development Bank. Non-sovereign Guaranteed Operations.	*Infrastructure*, **energy,** transport, water and sanitation, industry, agribusiness, natural resources, financial institutions, capital markets, trade finance, health care, education, tourism, corporate governance, corporate social responsibility, and **climate change.**
International Finance Corporation – World Bank Group	Frontier markets and International Development Association countries, sustainability, *infrastructure*, agribusiness, health and education, financial markets, and small and medium-sized enterprises.
Multilateral Investment Guarantee Agency – World Bank Group	International Development Association and conflict-affected countries, *infrastructure*, and South-South investment projects.

The table shows a clear focus among most IFIs on clean and renewable energy plus climate action

Source: International Finance Corporation, 2011

If private companies sponsor a geothermal project in a developing or emerging country, or form a consortium with investors from other countries to sponsor a project, an IFI such as the World Bank could be an ideal partner to help mobilize funds. Guarantees to facilitate participation by private investment banks could also be critical.

IFIs, which are also development institutions, generally need to demonstrate that their financing is essential, beyond what commercial finance would provide on its own, and that they can add value through risk mitigation and improved project design that leads to better overall development outcomes. They need to ensure that they crowd in investment and do not harm development of private financial markets. Most IFIs recognize this need, and many call their special role "additionality," that is, the value they bring to a project beyond what private sector financial institutions could typically offer (IFC, 2011).

IFI participation can help projects in developing and emerging countries in at least two notable ways: (1) making them more commercially viable through, for example, better finance, improved risk mitigation, advice; and (2) improving their developmental outcomes by, for example, providing the advice and standard setting that lead to better operations, products, and services; stronger environmental, social, and corporate governance; or projects that are more inclusive (IFC, 2011). IFIs also tend to provide finance with longer maturities, which is generally beyond the risk appetite of private capital (IFC, 2011).

IFIs clearly can be catalysts to support PPP projects, including those in the energy sector; however, they tend to be bureaucratic, which can cause problems and be costly for private sector partners. Nam Theun 2, discussed in this book, is an example of a project with long and costly preparation time. Other issues can also arise such as instances of moral hazard, also discussed in this book, including the case of Enron in Indonesia.

The Need for IFIs to Make More Use of Their Guarantee Powers in Developing and Emerging Countries

The ongoing debate about the role of IFIs increasingly recognizes the importance of making greater use of the risk mitigation potential inherent in their unique multilateral structure (Asian Development Bank, 2006). This discussion surfaces from time to time both within IFIs and elsewhere. About ten years ago the World Economic Forum[40], for example, argued strongly for IFIs

[40] The World Economic Forum's Financing for Development Initiative comprises more than 200 global experts from financial institutions, corporations, governments, international organizations, universities, and nongovernmental organizations, who offer their views on

to put risk mitigation instruments to better use in order to attract increased commercial investment in development projects. The World Economic Forum issued a comprehensive report entitled "Building on the Monterrey Consensus: The Untapped Potential of Development Finance Institutions to Catalyze Private Investment". In this report the World Economic Forum specifically asserted that: "…the weight of [development finance institution] activities should shift over time from direct lending to facilitating the mobilization of resources from the world's large private savings pools – international and domestic – for development-oriented investment through: wider use of risk mitigation instruments to alleviate part of the risk faced by investors; and stronger direct support for capacity building to strengthen the enabling environment for investment" (World Economic Forum, 2006, p. 9).

Furthermore, the World Economic Forum argued that development financial institutions should "…adapt their services, culture and capital allocation to the imperative of 'crowding in' domestic and foreign private investment by placing much more emphasis on such risk mitigation instruments as partial guarantees as transitional strategy and on capacity building…" (World Economic Forum 2006, p. 10) and that "an international consensus has emerged, embodied by the Monterrey Consensus, that a deeper partnership between the public and private sector is needed if we are to achieve common development objectives" (World Economic Forum, 2006, p. 10). In its final recommendations the World Economic Forum says: "The overwhelming majority of expert participants in the project recommended a major expansion of risk mitigation activity by [development finance institutions] …" (World Economic Forum, 2006, p. 15).

The World Economic Forum thus sent a crystal clear signal to the IFIs. In its report entitled "Review of ADB's Credit Enhancement Operations" the Asian Development Bank took note of the World Economic Forum's view and referred to their 2006 report several times (Asian Development Bank, 2006).

While a clear need exists for risk mitigation in developing and emerging countries for sectors such as clean energy, it seems that the World Bank Group as well as regional IFIs have some way to go to make those instruments widely used. IFIs need to do a better job in coordinating risk mitigation activities within themselves and spend more effort in marketing those products and to make them more efficient and more cost effective for the private sector and shorten their processing time. IFIs also need to help ensure that the deal between the public

---

improving the effectiveness of efforts to stimulate private sector investment in developing countries.

and the private sector is fair and help address moral hazard problems that can be associated with those instruments. This will be discussed in more detail later in this book in the chapter on public private partnerships – costs and benefits.

Chapter 5

THE WORLD BANK GROUP AND THE EFFECTIVENESS OF ITS RISK MITIGATION INSTRUMENTS

The effectiveness of the risk mitigation instruments offered by IFIs and the performance of those institutions require constant review and scrutiny. About seven years ago the World Bank Group issued a comprehensive assessment in a report entitled: "The World Bank Group Guarantee Instruments 1990-2007. An Independent Evaluation" (World Bank, 2009). Unfortunately, no such comprehensive evaluation has been done since then.

As part of this evaluation, in 2008 the Independent Evaluation Group[41] at the World Bank conducted a survey to solicit views among its staff about the use and effectiveness of guarantee instruments (World Bank, 2009). A survey questionnaire was sent to 363 staff, of whom 206 responded. Among the information that the survey revealed is that World Bank Group staff are familiar with their own products, but not with the guarantee products of other World Bank Group institutions. For example, only one-fifth of the staff of the International Finance Corporation (IFC)[42] were familiar with the products of the International Bank for Reconstruction and Development (IBRD) or those of the

[41] The Independent Evaluation Group is an independent agency within the World Bank Group whose purpose is to assess the relevance and impact of the Bank Group's work to reduce poverty and improve people's lives in a sustainable manner. The Independent Evaluation Group reports directly to the executive directors. Independent Evaluation Group goals are to contribute to the objective basis for assessing results, to provide accountability in the achievement of development objectives, and to share learning gained from experience. The Independent Evaluation Group makes its findings available to the broader development community and the public at large (World Bank, 2015a).

[42] The International Finance Corporation (IFC) is the private sector arm of the World Bank Group.

International Development Association (IDA).[43] In fact, IFC staff were also not familiar with the products the Multilateral Investment Guarantee Agency (MIGA).

According to the survey, more than 85 percent of World Bank Group staff felt that the most critical benefits of World Bank Group guarantee instruments were enhanced image of financial soundness and improved rates and tenors. Other benefits included e.g., the World Bank Group's role as an honest broker and securing other investors (World Bank, 2009).

It is also notable how few guarantees and insurances had been issued from an institution as large as the World Bank Group. A high proportion of staff felt that changes are needed to improve the World Bank Group's guarantee instruments (World Bank, 2009). Interestingly enough, most World Bank Group staff felt that reducing the time and cost of processing guarantees and improving marketing were important for improving World Bank Group guarantee instruments. Furthermore, staff reported that clients proceeding with a project without a guarantee and long processing time were the main reason for dropped guarantee projects. Eighty percent of IFC staff reported that droppages occurred because the cost of the guarantee was too high for the client (World Bank, 2009).

IBRD, IDA and MIGA staff reported that project sponsors/investors most frequently originated a request for guarantees. IFC staff reported that host governments and staff of another World Bank Group institution are least likely to originate IFC guarantees.

On May 7, 2008 the Committee on Development Effectiveness at the World Bank considered the Independent Evaluation Group evaluation. Several speakers called for greater collaboration among World Bank Group institutions based on their comparative advantages, and strengthening the coherence of the products offered, including pricing. They also called for more coordinated World Bank Group efforts for marketing, increased staff knowledge of guarantee products, and appropriate staff incentives (World Bank, 2009, p. xxviii). Comments were also made about the need for the World Bank Group to think about a "single Window" for guarantee products (World Bank, 2009, p. xxvi).

As already mentioned, no evaluation by the Independent Evaluation Group has taken place at the World Bank since 2009, [44] making it difficult to assess whether the effectiveness of World Bank Group guarantee instruments has improved since 2009.

---

[43] The IBRD and the IDA are the public sector arms of the World Bank Group.

[44] According to an email from Jayasankar Shivakumar and Pankaj Gupta, World Bank Group, dated July 13. 2013.

Table 6 below provides information about World Bank Group guarantee products by institution, risk coverage, eligible borrower, project time, beneficiary, as well as pricing and indemnity requirements.

ENHANCING THE WORLD BANK'S OPERATIONAL POLICY FRAMEWORK ON GUARANTEES

The use of guarantees continues to be discussed within the World Bank Group. Recently the World Bank, for example, issued a report entitled: "Enhancing the World Bank's Operational Policy Framework on Guarantees" (World Bank, 2013c). In this report the World Bank recognizes that "Bank guarantees can help member countries mobilize private resources, but they are underused, partly due to policy constraints" (World Bank, 2013c, p. i). According to the World Bank, member countries' financing needs are large and growing. The bank notes that over the past decade, through IDA partial risk guarantees, a single US dollar (US$1) of IDA commitment leveraged, on average, almost US$6 of private capital and US$9 of total project financing. According to the report, World Bank guarantees thus provide a way to leverage resources when the scale and cost of transformational projects and programs would otherwise constrain the World Bank's participation (World Bank, 2013c).

According to the World Bank, the proposed reform of guarantees is part of management efforts aimed at enabling leverage of World Bank Group resources and synergies.[45] More flexible and accessible World Bank guarantees would

[45] According to the World Bank the reform has five broad features: *First*, opening access for all International Development Association (IDA) members to all types of World Bank guarantees under certain fiscal considerations. *Second*, lifting undue limitations on the use of World Bank guarantees. *Third*, fully incorporating World Bank guarantees into investment project financing and development policy financing. World Bank guarantees are currently viewed as a separate instrument. However, World Bank guarantees can be viewed as an alternative financing modality to World Bank loans since the underlying projects or policy and institutional reforms for which the World Bank provides guarantees are substantially similar to those supported by World Bank loans under investment project financing and development policy financing. Incorporating World Bank guarantees into investment project financing and development policy financing operational policies would enhance the visibility of this form of financing, facilitate use by World Bank teams and demand from client countries, and lift various constraints that arise from having separate policies. *Fourth*, clarifying aspects of the policy framework that were perceived as gaps or unclear, and were inhibiting the use of this form of financing. And *fifth*, modernizing World Bank guarantee policy in certain respects to help ensure that the World Bank's guarantee products offer the flexibility necessary to meet the changing and future needs of member countries (World Bank, 2013c, p. ii).

Table 6. World Bank Group Guarantee Products

Institution	*Risk coverage*	*Eligible borrower*	*Project type*	*Beneficiary*	*Pricing and indemnity requirements*
World Bank ***International Bank for Reconstruction and Development (IBRD)*** ***and*** ***International Development Association (IDA)***	**Credit Risk:** Default on payment obligations on bonds, loans, and non-loan related government obligations regardless of cause of default	• **For Development Policy Financing:** Sovereign governments • **For Investment Policy Financing:** Sovereign governments and sub-national entities such as states, municipalities and state-owned enterprises • **For non-loan related government obligations in relation to Investment Policy Financing**: Private and foreign public entities	• **For Development Policy Financing:** Public sector program of policy and institutional reforms • **For Investment Policy Financing**: Public sector projects • **For non-loan related government obligations in relation to Investment Policy Financing**: Private sector projects and public-to-public cross-border transactions • **For IBRD Enclave Investment Project Financing:** a. Project with strong financial flows in local currency in an IDA-only country that has sufficient alternative non-project related foreign exchange to cover the country's obligations to IBRD;	• **For Development Policy Financing**: Private lenders and bond investors • **For Investment Policy Financing:** Private lenders and bond investors • **For non-loan related government obligations in relation to Investment Policy Financing**: Private and foreign public entities	Priced similar to IBRD loans or IDA credits Government must sign indemnity agreement with the World Bank

Institution	*Risk coverage*	*Eligible borrower*	*Project type*	*Beneficiary*	*Pricing and indemnity requirements*
World Bank ***International Bank for Reconstruction and Development (IBRD)*** ***and*** ***International Development Association (IDA)***			b. Export-oriented, foreign exchange-generating project with private sector participation in IDA-only country		Priced similar to IBRD loans or IDA credits Government must sign indemnity agreement with the World Bank
	Political Risk: Default on payment obligations on loans caused by failure of government to meet specific performance obligations, including but not limited to: - Contractual performance of public counterparties - Change of laws and regulations - Expropriation - Non-convertibility and transferability of currency – War and civil disturbance, etc.	Private implementing entity in public private partnership (PPP) projects	• Projects with private sector participation such as build operate-transfer and concession projects; public private participation (PPP) projects; privatizations • **For IBRD Enclave Investment Project Financing:** a. Project with strong financial flows in local currency in an IDA-only country that has sufficient alternative non-project related foreign exchange to cover the country's obligations to IBRD b. Export-oriented, foreign exchange-generating project with private sector participation in IDA-only country	Private entity	

Table 6. (Continued)

Institution	*Risk coverage*	*Eligible borrower*	*Project type*	*Beneficiary*	*Pricing and indemnity requirements*
International Finance Corporation (IFC)	**Credit Risk:** Default on payment obligations on bonds, loans, asset portfolios, derivative contracts, trade finance, and other financial instruments	Private sector entity	Corporate, project, structured and trade financing	Private lenders, bond investors and other providers of credit to the private sector	Priced on a commercial basis No indemnity agreement required
Institution	*Risk coverage*	*Eligible borrower*	*Project type*	*Beneficiary*	*Pricing and indemnity requirements*
Multilateral Investment Guarantee Agency (MIGA)	**Credit Risk:** Default on payment obligations on any form of private commercial debt (non-honoring of financial obligations)	Sovereign or sub-sovereign government or state-owned enterprise	Public sector project involving cross-border financing	Commercial lenders providing credit to the public sector	Priced on a commercial basis Requires host country approval, but no indemnity agreement
	Political Risk: Losses on equity/quasi-equity or other forms of direct investment caused by: - Expropriation - Breach of contract - Currency inconvertibility and transfer restriction - War and civil disturbance	Private sector entity or government-owned company operating on a commercial basis	Private sector project involving cross-border investment	Private investor or lender or government-owned company operating on a commercial basis	

Source: World Bank n.d. Constructed by the author.

help the World Bank Group (IBRD/IDA, IFC and MIGA) to work together more effectively, tackling clients' needs and catalyzing private sector participation in pursuing member countries' objectives (World Bank, 2013c).

An example of natural convergence for World Bank Group support through guarantees comes in support to PPPs, where IFC loans, MIGA political risk insurance and World Bank guarantees have been deployed together. In practice, IFC guarantees are designed so that very little overlap occurs with MIGA and World Bank guarantees (World Bank, 2013c, p. 5). An example of this is the Nam Theun 2 hydropower project in Lao as well as the Bujagali hydropower project in Uganda discussed in a later chapter in this book. Financing and knowledge instruments provided across the World Bank Group to such projects can be complementary, optimizing use of the World Bank Group balance sheet.

As noted above, World Bank guarantees can be critical in helping countries mobilize private financing by sharing with private lenders the risk of debt service default or the occurrence of other specified risks that may cause debt service default due to a government or government owned entity's failure to fulfill its obligations. World Bank guarantees cover risks only to the extent necessary to obtain the required private financing. The guarantees are partial in nature (less than 100 percent) given that they are aimed at leveraging private financing. All World Bank guarantees require a sovereign counter-guarantee and indemnity, comparable to the requirement of a sovereign guarantee for World Bank lending to sub-sovereign and non-sovereign borrowers[46] (World Bank, 2013c, p. 5).

World Bank guarantees are distinct from MIGA and IFC guarantees in mobilizing private sector financing. Given that IBRD and IDA clients are first and foremost member countries, World Bank guarantees play a different yet complementary role to that of MIGA and IFC guarantees (see Table 7 below).

MIGA provides guarantees in the form of political risk insurance for cross-border direct investments for a wide range of private sector clients. IFC provides credit guarantees for private sector participants as their primary clients. Neither MIGA nor IFC guarantees require an explicit sovereign counter-guarantee. World Bank guarantees can also support private sector projects but only by backstopping public sector obligations for which the member country is willing to provide an indemnity (World Bank, 2013c, p. 5). See Table 8 below summarizing MIGA, IFC and World Bank (IBRD and IDA) guarantee pricing.

[46] IBRD Articles of Agreement, Article III, Section 4(i).

Table 7. World Bank Group. Main Guarantee/Insurance Products

Multilateral Investment Guarantee Agency (MIGA)	International Finance Corporation (IFC)	World Bank: International Bank for Reconstruction and Development (IBRD) International Development Association (IDA)
Political Risk Insurance • Currency inconvertibility and transfer restrictions • Expropriation •War, terrorism and civil Disturbance • Breach of contract • Non-honoring of sovereign financial obligations	• **Direct Debt Substitute:** credit guarantees (partial, full and credit-linked), risk participations and risk sharing facilities covering credit risk on private borrowers • **Commercial Operation**: Credit enhancement guarantees and risk sharing facilities covering commercial risks or combination of commercial and credit risks (e.g., guarantees of performance bonds) • **Global Trade Finance Program:** Coverage of non-payment by banks in markets for trade-related transactions • **Guaranteed Offshore Liquidity Facility:** Coverage of currency transfer and convertibility risk • **Client Risk Management Guarantees:** Coverage of counterparty credit risk on private borrowers in derivatives transactions • **B-loan program** Implicitly offers risk mitigation similar to MIGA's transfer restriction	• **Partial Risk Guarantee** Covers lenders against debt service default, normally for a private sector project, caused by a government's failure to meet specific obligations • **Partial Credit Guarantee** Covers lenders against debt service default, typically in a public sector project • **Policy-Based Guarantee** Covers lenders against debt service default under a sovereign borrowing made in support of a program of policy and institutional reforms

Source: (World Bank, 2013c)

Table 8. International Bank for Reconstruction and Development (IBRD) and International Development Association (IDA) Guarantee Pricing

	Fee Type	IBRD			IDA
		Partial Risk Guarantees	Partial Risk Guarantee Enclave for IDA countries	Partial Credit Guarantees and Policy-based Guarantees	Partial Risk Guarantees
Upfront Charges (one-time fees)	Front-End Fee	25 bps[47] (on max. exposure under the guarantee)			N/A
	Initiation Fee (1)	15 bps on guaranteed amount or USD100,000 (whichever is greater)		N/A	15 bps on guaranteed amount or US$100,000 (whichever is greater)
	Processing Fee (1, 2)	Up to 50 bps of guaranteed amount		N/A	Up to 50 bps of guaranteed amount
Recurring Charges	Guarantee Fee (3)				
	Average Maturity Up to 12 years	50 bps	200 bps	50 bps	75 bps
	Average Maturity greater than 12 to 15 years	60 bps	210 bps	60 bps	75 bps
	Average Maturity greater than 15 to 18 years	70 bps	220 bps	70 bps	75 bps
	Standby Fee	N/A			0 bps (4)

1. For all private borrowers, i.e., only applicable to partial risk guarantees.
2. Determined on a case by case basis. Exceptional projects can be charged over 50 basis points (bps) of the guaranteed amount.
3. The guarantee fee is charged on: the maximum aggregate disbursed and outstanding amount for partial risk guarantees, and on the present value of guarantee exposure for partial credit guarantees and policy-based guarantees. For IBRD guarantees, the guarantee fee includes an annual maturity premium of 0.00% for maturities up 12 years, 0.10% for maturities greater than 12 and up to 15 years, and 0.20% for maturities greater than 15 and up to 18 years.
4. Equivalent to the commitment charge on IDA Credits in financial year 2010.

Source: World Bank n.d. [48]

---

[47] Basis points i.e., bps.

[48] See further http://siteresources.worldbank.org/INTGUARANTEES/Resources/guaranteepricingFY11.pdf.

Currently the World Bank offers three types of guarantees. Partial risk guarantees cover private lenders against debt service defaults on loans, normally for a private sector project, when such defaults are caused by a government's failure to meet specific obligations under project contracts to which it is a party. Partial credit guarantees cover private lenders against debt service defaults on a specified portion of loans, normally for a public sector project, irrespective of the cause of such debt service default. Policy-based guarantees cover private lenders against debt service defaults under a sovereign borrowing made in support of a program of policy and institutional reforms (World Bank, 2013c, p. 5).

To meet growing infrastructure needs where project feasibility is not only constrained by financing but also by payment risk mitigation for project outputs or for government termination payment obligations, the World Bank has, for example, offered partial risk guarantees (World Bank, 2013c, p. 8).

In order to avoid entangling the World Bank in the substance of any contractual dispute between the parties, World Bank guarantee policy requires (a) that the underlying contracts for partial risk guarantees contain appropriate dispute resolution procedures; and (b) that if a dispute arises about the government's obligations, the World Bank guarantee is triggered only after the government's liability has been determined in accordance with such procedures. When the World Bank partial risk guarantee indirectly covers regular government off-take payments (as can be the case with geothermal and hydropower projects), the government is usually liable for timely payments of the full amount invoiced, even when an underlying dispute is ongoing. In these cases there has been considerable demand for World Bank partial risk guarantees to backstop the entire government payment obligation, in respect of which the World Bank would obtain reimbursement under the indemnity provided by the country (World Bank, 2013c, p. 17). It is questionable whether this is fair treatment of host governments as the arrangement reduces incentives for the private sector to renegotiate projects with host governments during economic and financial crises, as will be discussed in a later chapter in this book.

The signs now are that the World Bank Group is increasing its use of guarantee instruments to mobilize private sector capital for various projects in developing and emerging market economies. These include projects on renewable energy development, hydropower and energy efficiency, see Table 9 below. As the table shows, guarantees can be effective in mobilizing private capital.

Table 9. Guarantee transactions 2013-2015. Further details about each of these World Bank Group board approved projects can be found on the World Bank's external website

Country	Project Name	Guarantee Amount US$ million	Private Capital Mobilized US$ million
2013			
Croatia	Guarantee for support of Financial Intermediary (Croatian Bank for Reconstruction and Development)	40	50
Cote D'Ivoire	CI - 27 Gas Field Expansion	60	960
Macedonia	Second Policy Based Guarantee for the Financial Sector	202	325
2014			
Pakistan	Dasu Hydropower Stage 1	460	2446
Mali, Mauritania, Senegal	Banda Gas to Power Project (sub-projects for each country)	261	650
Nigeria	Azura: Nigeria Power Sector Independent Power Producer Series	245	573
Nigeria	Exxon: Nigeria Power Sector Independent Power Producer Series	150	1136
Senegal	Taiba Ndiaye Thermal Power Plant	40	54
Uganda	Renewable Energy Development Program	160	500
Maldives	Solar PV Program Support (Scaling-Up Renewable Energy in Low Income Countries ASPIRE Program)	16	42
2015			
Ghana	Ghana Macroeconomic Stability Policy Based Guarantee	400	1,000
Angola	Angola Fiscal Management Development Policy Financing	200	200
Kenya	Kenya Electricity Modernization Program	200	500
India	The Partial Risk Sharing Facility for Energy Efficiency	25	127

Source: World Bank Group, 2015[49]

[49] According to an email dated 23 July 2015 from Katharine C. Baragona, Senior Infrastructure Finance Specialist, Project Finance and Guarantees, Energy & Extractives Global Practice, the World Bank Group.

Chapter 6

GEOTHERMAL DEVELOPMENT AND UPFRONT DEVELOPMENT COSTS

In addition to the risks associated with cross border investment and operations that affect both geothermal and hydro power projects, geothermal projects also suffer from risks not found in other thermal power generation projects, most notably higher upfront development costs associated with uncertainty as to site capacity[50] (Delmon, 2009). Risks associated with geothermal projects are high during the pre-survey, exploration and test drilling phases.[51] Validating geothermal resources through test drilling is capital intensive. Commercial financing for test drilling is generally hard to find. For this reason, private equity, government support or international grants are often the only sources of capital to undertake test drilling. These risks are not specific to the developing and emerging markets that this book focuses on, but it is often more challenging to obtain both private and public capital in these markets than in high income countries. Private sector investors are less likely to engage in the uncertain business environment of developing and emerging market economies; moreover, governments in those countries do not have as strong a tax base as do high income developed countries. Because of this, bilateral and multilateral donors who can provide grants have an important role to play.

---

[50] Even if capacity exists initially, it may not be sustainable over the lifetime of the project.

[51] Until the first borehole has been drilled into the geothermal reservoir, developers cannot be sure about the exact parameters (temperature and flow rate) of a planned geothermal electricity project. Once drilling has taken place, *in situ* pump tests, temperature and hydrological measurements then reduce the resource risk and can make it easier to attract external capital (GEOELEC, 2013, p. 6).

Over the last three decades, IFIs have done surprisingly little to mobilize funding for resource risk mitigation for geothermal projects in developing and emerging countries (see Table 10). They have mostly spent their resources on the less risky production phase after the riskier exploratory phase is completed. According to the World Bank, about three billion US$ have gone into geothermal investments of which only US$124 million were used for the exploration phase and US$62 million on technical assistance, see Table 10. Thus 94 percent has been spent on above-the-ground infrastructure compared with only four percent on exploration and two percent on technical assistance. This is unacceptable and quite the opposite to what one would expect given the development role the IFIs discussed in this book are supposed to play. Here IFIs have been failing in their role to promote geothermal development in developing and emerging countries.

Table 10. Three decades of cumulative lending for geothermal energy development (US$ million)

International financial institutions (IFIs)	Exploratory Phase	Production Phase	Technical Assistance	Total
World Bank	117	1,544	48	1,710
African Development Bank	4	124	-	129
Asian Development Bank	-	554	3	557
European Investment Bank	-	256	-	256
Inter-American Development Bank	3	403	11	416
Total	124	2,881	62	3,068

Source: World Bank, 2013b

Scaling up geothermal development by addressing the resource risk through sustained international effort has been discussed at the World Bank, including raising US$500 million for exploration and drilling activity under the newly established Global Geothermal Development Plan.[52] This is a modest amount,

[52] On 6 March 2013, World Bank Managing Director Sri Mulyani Intrawadi launched the Global Geothermal Development Plan in Reykjavik, Iceland. This initiative is a call to all donors and sponsors of geothermal energy to join hands in mobilizing additional concessional resources and fund a pipeline of geothermal resource validation projects. According to the World Bank this is a way to address a critical financing gap and to help scale up geothermal energy and

given global needs, but can nevertheless be considered an important step in the right direction. According to the World Bank US$250 million has been mobilized. Table 11 below lists current activities under the Global Geothermal Development Plan that are located in Africa, Asia and Latin America. In Europe a European Geothermal Risk Insurance Fund is being prepared (GEOELEC, 2013).

Table 11. The Global Geothermal Development Plan. Current Activities

Country	Scope
Latin America and the Caribbean Region	Partnership with KfW Development Bank Geothermal Development Facility
Chile	Co-financing for Clean Technology Fund Technical Assistance component on Sustainable Geothermal Development project
Nicaragua	Technical assistance to IDA lending operation identification and preparation (w/IFC)
Dominica	Technical assistance to IDA lending operation identification and preparation (w/Agence Française de Développement)
St Lucia	Technical assistance and co-financing of Small Island Developing States- DOCK/Global Environmental Facility technical assistance operation
Armenia	Technical assistance to prepare Scaling-Up Renewable Energy in Low Income Countries Program financed Geothermal Exploratory Drilling Project
Turkey	Technical assistance to IBRD-Clean Technology Fund Geothermal Development Project preparation
Africa Region	Partnership with Iceland International Development Agency-Nordic Development Fund Geothermal Exploration Project
Africa Region	Approach Paper - Scaling Up Geothermal Power Generation in Africa
Djibouti	Co-financing for IDA (and other donors) Geothermal Power Generation Project
Ethiopia	Peer review support to IFC implemented, SREP funded Geothermal development strategy
Tanzania	Capacity building support for Geothermal Development Company (TBD), with Netherlands, African Development Bank
Indonesia	Technical assistance to IBRD-Clean Technology Fund Geothermal Development Project preparation
Fiji	Technical assistance support to identification and preparation of geothermal project (identification)

Source: World Bank, 2015b

expand clean energy access (According to an email from the World Bank on 17 February 2014).

The European Bank for Reconstruction and Development (EBRD) is launching a US$125 million framework to support early stage geothermal development in Turkey. The project will combine up to US$100 million from EBRD funds with US$25 million from the Clean Technology Fund.

According to the EBRD the project will provide loans to private sector investors aimed at bridging the funding gap existing at the early stage of development of geothermal power plants. In particular, the Clean Technology Fund will be supporting sub-project sponsors during the early stage of geothermal power plant development through a risk mitigation financial instrument. The facility is also intended to provide technical assistance to promote best practice in the development of geothermal resources for power and heat generation, particularly at the early stage. This will include supporting developers during their exploration and production drilling campaigns to minimize technical and financial risks.

The geothermal potential in Turkey is estimated by EBRD to be around four gigawatts, of which only 300 megawatts has been developed. One of the key barriers to development is a combination of (i) high specific investment costs, with (ii) high project development risks, particularly in the initial phase.[53]

IFIs have so far done little to help fund the high upfront development costs for geothermal projects in developing and emerging countries. Instead they have primarily invested in the least risky production phase. This is unacceptable and IFIs should team up with other development institutions with grants and concessional lending to help make geothermal projects possible. This is especially urgent in the poorest developing countries where possibilities for government support is limited and where private sector engagement is less likely than in more advanced economies.

[53] For more detail see, http://www.ebrd.com/work-with-us/projects/psd/early-stage-geothermal-support-framework.html.

Chapter 7

PUBLIC PRIVATE PARTNERSHIPS – COSTS AND BENEFITS

Clean energy investments such as geothermal and hydropower projects tend to be large, capital intensive and with long repayment periods. These projects are challenging, especially in developing and emerging market economies that are in transition and sometimes characterized by a changing and unpredictable political situation and unfavorable business environments.

For geothermal projects, several models can be used to mobilize capital and cover the resource risks. These include, for example: (i) an enterprise in public ownership, (ii) a private company developer and (iii) a public private partnership (PPP). In the first model, a public sector company would typically undertake the whole project, including the resource risks. This could be a combination of national and municipal authorities such as in the case of Iceland. In the second model a private sector developer might be a large company with a strong balance sheet as in the case of Chevron in the Philippines. In this case the private company has the financial capacity to undertake the whole project. In the third model the public and private sectors work in partnership, i.e., a public private partnership (PPP).

PPPs enable pooling of public, private and donor funds for clean energy investment in developing and emerging market economies. Those economies are also eligible for support from international financial institutions. A well designed PPP can be a venue for scaling up funding for clean energy internationally and thus contribute to the battle against climate change. Thus a potential global benefit exists in using this arrangement. The focus of this chapter is on PPPs.

But what exactly is a public private partnership? Many different definitions are used for PPPs by different individuals and institutions. One definition is "any public sector service provided partially or wholly by the private sector" (Delmon, 2009, p. 601). Another definition is "co-operative institutional arrangements between public and private sector actors" (Hodge and Greve, 2009, p. 33). The World Bank has defined PPPs as "the transfer to the private sector of investment projects that traditionally have been executed or financed by the public sector" (World Bank, 2008, p. 93). The Organization for Economic Co-operation and Development has defined PPPs as "long term contractual arrangements between the government and a private partner whereby the latter delivers and funds public services using a capital asset, sharing the associated risks" (OECD 2012, p.18). Notably, the Organization for Economic Co-operation and Development and World Bank definitions mention capital investment by the private sector. The other definitions are more general, highlighting service and cooperation.

Potential Benefits from Using PPPs in Developing and Emerging Economies

Why would governments of developing and emerging market economies want to cooperate with the private sector under a PPP arrangement? The answer is: for several reasons. A PPP can be a feasible option for governments because in most developing and many emerging countries the government has limited capacity to mobilize funds. This means that a partnership with the private sector can help the government obtain much-needed long term capital. The private sector may also possess technical knowledge that the public sector does not have and be a more efficient operator of power plants. In this case a public private partnership can be formed to allow for a sharing of the risks and rewards from the project.

PPP can be a favored model in developing and emerging markets and be beneficial if well designed and if risks and rewards are shared fairly. However, little point exists in forming PPPs if, for example, the private sector partner captures most or all the benefits, or if the government keeps changing the rules of the game resulting in a non-viable project.

With pressure on physical infrastructure and limited resources, governments of developing and emerging countries may want to cooperate with the private sector via a PPP to help finance, build, and/or operate public projects.

While doing so, the government could at least in theory: (1) utilize the better skills and better management from the private sector that may lead to increased efficiency for the project in a competitive environment, (2) access private sector funds to undertake more projects than would be possible with public funds alone – this can contribute to fiscal stabilization, and increase investment and growth, (3) provide more affordable and better services to end-users, and (4) share risks with the private sector (see, for example, Leruth, 2009; de Palma et al., 2009; Estache, 2005).

Developing and emerging economies on the one hand and high-income economies on the other may have different reasons for participating in public private partnerships. As Hart points out, "Policy makers frequently argue that PPPs are good because the private sector is a cheaper source of financing or insurance than the public sector." Furthermore, he emphasizes that "This thinking is strange for an economist since it is hard to imagine an agent that is more able to borrow or to provide insurance than the government (with its enormous powers of taxation)" (Hart, 2003, p. c75). Leruth also argues "the government is often able to borrow at [an] almost risk-free rate (no credit risk), which gives it an advantage" (Leruth, 2009, p. 230).

These arguments may well be true in countries that enjoy strong creditworthiness (e.g., via AAA or other high credit rating status). The countries discussed in this book, however, are developing and emerging countries. They often have large unutilized energy resources and strong medium- or long-term demand for energy, but their creditworthiness is limited. Their nationals often have limited ability to pay for the services rendered to them and the government has weak capacity to force them to do so through taxation. Such governments can be risky partners for the private sector in a PPP. In this situation, efficient and effective risk allocation is the key to success, and the international community can play a constructive role, for example, through participation by IFIs, which can involve a variety of funding and mitigation instruments.

When discussing green infrastructure finance, the World Bank states that "the international community has recognized that the majority of new investment financing will need to come from private sources" (World Bank, 2012a, p. 16). The World Bank's focus is on developing and emerging markets. This means that some sort of public private partnerships will need to be formed for a large share of clean energy investment in those markets in transition. This is a major challenge not only for the private sector, but also for participating developing and emerging countries. Moreover, as the World Bank has stated, "developing a framework for improved collaboration between public and private sectors could greatly benefit green infrastructure financing mechanisms" (World Bank, 2012a, p. 6).

Interest in PPPs is growing among Asian countries, including a notably prominent market in China. The Association of Southeast Asian Nations (ASEAN) for example recently stated that "PPPs are seen to be beneficial in meeting ASEAN infrastructure needs, estimated by the Asian Development Bank at USD 60 billion per annum. Private participation in infrastructure provision can enhance existing public capacity in providing economic (e.g., transport, telecommunication, power, water and sanitation) and social (e.g., health and education) infrastructures" (ASEAN, 2014). Great need is also present in other developing and emerging regions such as Africa and Latin America.

CHALLENGES TO USING PPPS IN DEVELOPING AND EMERGING ECONOMIES

According to the Public-Private Infrastructure Advisory Facility54 countries using the PPP model need to have: strong institutions, legal systems and rule of law, high standards of public and corporate governance, transparency, competition, protection of investments, enforcement of laws, and dispute resolution mechanisms (World Bank, 2015a). This is an impressive list, but some would argue that it is unrealistic even for developed high income countries. If international financial institutions insisted that those criteria be satisfied before an investment could take place using the PPP model, very few

[54] The Public-Private Infrastructure Advisory Facility was created in 1999 to act as a catalyst to increase private sector participation in emerging markets. It provides technical assistance to governments to support creation of a sound enabling environment for the provision of basic infrastructure services by the private sector. According to the Public-Private Infrastructure Advisory Facility a sound business-enabling environment consists of strong institutions, legal systems and rule of law, high standards of public and corporate governance, transparency, competition, protection of investments, enforcement of laws, and dispute resolution mechanisms. The Public-Private Infrastructure Advisory Facility is a multi-donor technical assistance facility, financed by 17 multilateral and bilateral donors: the Asian Development Bank, Australia, Austria, Canada, the European Bank for Reconstruction and Development, France, Germany, the International Finance Corporation, Italy, Japan, the Millennium Challenge Corporation, the Netherlands, Sweden, Switzerland, the United Kingdom, the United States, and the World Bank. Public-Private Infrastructure Advisory Facility funds are untied and grants are provided on a demand-driven basis, see further: http://www.ppiaf.org/node/23#What%20is%20PPIAF?.

places in the world would receive any private finance. This would especially be a challenge in developing Africa.

Because of this challenge, key international organizations such as ASEAN, the European Union, the International Monetary Fund, the United Nations, and the World Bank have formulated and displayed substantial policy documents on PPPs. A recent paper on PPPs shows that international organizations use roughly the same categories and conceptions of stages for what is needed to establish effective PPPs (Greve, 2015).

Multilateral as well as bilateral financial organizations can at least in theory be a catalyst in supporting PPP projects, including in the clean energy sector. Institutions such as the World Bank Group and regional development banks can, via their policy dialogue, assist governments when undertaking reforms needed for effective use of the PPP model. IFI support can be important because developing and emerging countries often have limited capacity to negotiate with multinational enterprises. Since the Public-Private Infrastructure Advisory Facility is a multi-donor technical assistance facility, financed by 17 multilateral and bilateral donors, it can play an important role here.

But, as mentioned before, using the PPP model and working in partnership with IFIs and bilateral development institutions can also come at a cost. A recent book published by the World Bank, entitled "Public Private Partnerships in Europe and Central Asia – Designing Crisis-Resilient Strategies and Bankable Projects", comments very cautiously that "working with these institutions may also lengthen the project development process, given specific requirements in terms of environmental and social safeguards requirements and stringent procurement procedures" (Cuttaree and Mandri-Perrott, 2011, p. 59). Another recent book also published by the World Bank entitled: "Doing a dam better: the Lao People's Democratic Republic and the story of Nam Theun 2 (NT2)", is more critical when discussing World Bank cooperation with the private sector. The authors simply state that: "The bad news is that the World Bank is seen as a high-cost/high-hassle partner of last resort. There is therefore a critical need to reduce the costs the private sector incurs for doing business with the World Bank. Doing so will require the World Bank to better understand the constraints under which the private sector works" (Porter and Shivakumar, 2010, p. 22). These comments are especially notable given that the authors have both served as World Bank country directors. Shivakumar, for example, played an important role in enabling the Nam Theun 2 hydropower project in Lao, discussed is some detail later in this book.

Host Government Cooperation with the Private Sector, Moral Hazard and Times of Crisis

PPPs are not only a challenge for the private sector. Cooperating with private enterprise can also be a serious challenge for host governments in developing and emerging market economies, especially during times of economic and financial crisis. In their book, *Making Foreign Investment Safe – Property Rights and National Sovereignty*, Wells and Ahmed (2007) document a dramatic dispute between CalEnergy, an entity founded as a consulting and service company for geothermal power in North America, and the Indonesian government. This dispute resulted in a claim under official political risk insurance filed by CalEnergy and paid by the US government agency, the Overseas Private Investment Corporation. Less dramatic, but still consequential, was the dispute between Enron, then a private power developer, and the Indonesian government, that ended with the Multilateral Investment Guarantee Agency (MIGA), the insurance agency of the World Bank Group, paying out its first claim ever.

As Wells and Ahmed state; "The Indonesian experience with official political risk insurance has not been the only one that has made developing countries a bit wary. From the point of view of host countries, the new property rights system was not being very constructive. The moral hazard associated with the insurance surely had encouraged some investors to avoid renegotiation when economic crisis hit Indonesia and other countries. And Indonesians saw themselves as having few rights when CalEnergy filed a claim with [the Overseas Private Investment Corporation] or when [the Overseas Private Investment Corporation] sought reimbursement from Indonesia" (Wells and Ahmed, 2007, p. 246). Insurance can thus under certain conditions put the host government in a very difficult position vis-à-vis a private sector investor who is insured, in this case with the powerful US agency Overseas Private Investment Corporation. Moreover, Wells and Ahmed also criticize the public agencies providing guarantees as follows: "Further, and unlike many private insurers, public agencies appear to pay little attention to actual or potential moral hazards. If [the Overseas Private Investment Corporation] would insure only a smaller percentage of the equity, forcing more of the risk onto the investor, one aspect of moral hazard ought to decline. Second, allowing [the Overseas Private Investment Corporation] to seek something less than full reimbursement from host countries would reduce the perception that it faces no loss if it decides to pay claims" (Wells and Ahmed, 2007, p. 246).

In the case of the Enron dispute in Indonesia, MIGA paid Enron in the end. MIGA then demanded reimbursement from the government of Indonesia and got it. A good relationship with the World Bank Group, which MIGA is part of, must also have been considered important for the government of Indonesia because of other projects and programs that could be jeopardized in the event of an unresolved dispute. It is, however, questionable whether the World Bank Group should use its leverage in this way.

While the tensions between the government of Indonesia and MIGA did not run as high as in the case of the Overseas Private Investment Corporation, both were costly for the government of Indonesia in terms of money and international relations. In their concluding chapter, Wells and Ahmed raise doubts about whether the international system of property rights can accomplish its goal of encouraging foreign investment that is helpful for the host country. Nevertheless, and in spite of the serious problems that a host government can experience, they recognize the need for public private partnerships when they state that "Meeting the huge infrastructure needs for development will surely require a mix of private investment, state investment, multilateral lending, hybrid arrangements, and very substantial aid money" (Wells and Ahmed, 2007, p. 283).

IFIs are well placed to mitigate risks at competitive prices. As Wells says, "official insurance can be priced low, since the threat of sanctions by the organizations backing the insurance sharply reduce the chances of the events being insured against occurring" (Wells, 2005, p. 91).

Regional and global economic crises, such as those that struck Asia in 1997/98, Argentina in 2001, and the whole world in 2008, can pose significant threats to the international investment regime. As Salacus (2010) states, "Countries under great stress, faced with potential social and political upheaval as a result of rapidly declining standards of living, often seek radical solutions and are impatient with international investment rules that may restrict their latitude of action. For example, during times of economic crisis, governments may be unwilling to grant national treatment to foreign investors, to avoid changing regulations in the name of "fair and equitable treatment," and to refrain from seizing vital national resources held by foreigners simply because they have made treaty promises not to expropriate. Thus Argentina, to cope with one the most serious economic and financial crisis in its history in 2001–02, took a series of measures that foreign investors believed violated their legal rights and economic interests, resulting in the initiation of numerous investor-state arbitration claims" (Salacuse, 2010, p. 471). International financial institutions can play an important role in resolving disputes between host governments and foreign investors during economic and financial crises and they should make efforts to facilitate settlements that are fair for all parties involved. It should be a last resort action to pay out a claim to the private investor and then use the leverage of the IFI to force reimbursement from the host government.

PPPs - Allocating Risks and Sharing Rewards

In spite of the many potential problems and issues that can be associated with public-private cooperation and in operating public-private partnerships, PPPs can be a feasible platform to fund infrastructure development and to increase the efficiency of public sector service delivery in developing and emerging economies. The PPP becomes a venue for the public and private sectors to cooperate on a project that would traditionally have been in the public domain. Infrastructure projects in the clean energy sector are often large, capital intensive and long term. Repayment periods are also often long. It can take a private investor 10 to 30 years to recover the investment and project returns. It is often challenging to maintain a good working relationship within a PPP for such a long time, while the local, regional and global environment may change dramatically during this period.

The private sector has for years been recognized as a significant financing source for meeting developing country investment requirements. However, financial markets remain largely untapped for this purpose and have yet to live up to their potential (Asian Development Bank, 2006). PPPs are one platform worth considering for the private sector to engage in infrastructure projects, including geothermal and hydropower projects. Private capital, bilateral and multilateral donor support, including IFIs, and public funds can be combined in a PPP project. A well designed policy and institutional framework for PPPs offers the opportunity to leverage and combine all three sources of financing and expertise without crowding out private investment. By forming a PPP both public and private sectors can share the risks and rewards of infrastructure projects.

Private sector funding and participation in clean energy projects is a challenge for various reasons. Among them is that the host government is often the only buyer of the electricity or hot water produced i.e., it is the so called offtake purchaser[55]. As discussed earlier in this book, many developing and

[55] An offtake purchaser is a purchaser of the product produced by a project. In the case of a power project the product produced is the electricity generated.

emerging countries with large clean energy potential have limited creditworthiness. They have low per capita income and are often going through economic and political transition. Developing and emerging countries can also be vulnerable during regional and global crises. In such cases the sponsors[56] of a project may hesitate to inject funding because of uncertainty with the income stream from the investment. Lenders, including commercial investment banks, would also often hesitate to provide loans to such projects because of uncertainty whether the project company, whose income stream is at risk, can service its loans. Sponsors may also hesitate to provide equity capital.

To engage in cooperation, the public and private sectors can employ several different schemes[57] including the so called BOT, i.e., Build-Operate-Transfer (International Monetary Fund, 2004). In BOT projects the private sector is responsible for financing, constructing and operating the project. Under this arrangement the host country grants a concession[58], i.e., the right for a private firm to undertake a public sector project and operate it over an agreed period. When the concession expires the ownership of the project is transferred back to the party granting the concession.[59]

The partners typically involved in a BOT project are: the project company that undertakes the project, the host government (that can also be the offtake/power purchaser and guarantor), the shareholders, the lenders, the grantor, the construction contractor, the operator, the offtake purchaser/power purchaser, and the input supplier. Figure 2 below shows a typical PPP BOT contractual structure.

The project company uses the income stream from the project to service its debt and to pay returns to its investors.[60] The lenders to a BOT project might, for example, be commercial investment banks, IFIs and bilateral agencies. The IFIs could also serve as guarantors e.g., for payment to the lenders, including commercial investment banks. National institutions such as export credit agencies, which support trade finance (goods and services), can also play a

---

[56] A sponsor of a project is a party wishing to develop or undertake a project. A sponsor would normally provide financial support for the project e.g., early equity capital.

[57] PPP schemes and modalities other than Build-Operate-Transfer (BOT) include for example: Build-Own-Operate-Transfer (BOOT), Build-Rent-Own-Transfer (BROT), Build-Lease-Operate-Transfer (BLOT), Build-Transfer-Operate (BTO).

[58] A concession is the right granted by the host government for a private company to undertake a public sector project and operate it over an agreed period.

[59] For a comprehensive discussion on BOTs see, for example, Jeffrey Delmon's outstanding book on Private Sector Investment in Infrastructure (Delmon, 2009).

[60] i.e., equity contributors to the project company.

constructive role in reducing the risks taken by private sector investors see e.g., Dinh and Hilmarsson (2012a, 2012b and 2012c).

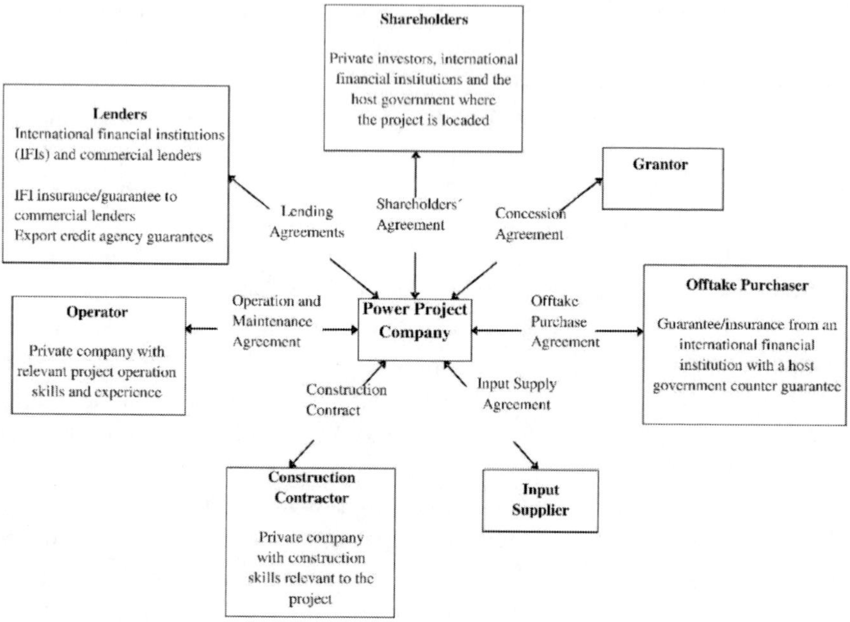

Source: Constructed by the author.

Figure 2. Typical PPP BOT contractual structure.

The lenders would typically be keen to manage their risks[61]and would receive a fixed margin on their loan whereas the shareholders[62] maximize the profits on their equity investment. In addition to obtaining funding for the project, the project company procures the design and coordinates the construction and operation of the project in linc with the requirements of the concession agreement. Project company shareholders often include firms with construction and operation experience, and with offtake purchase capabilities (Delmon, 2009, p. 98).

The offtake purchase agreement secures the project payment stream. The offtake purchaser will be looking for guaranteed long term output from the project. The credit risk associated with the offtake purchaser will be of particular

[61] i.e., they would only take measurable and measured risks.
[62] i.e., equity holders in the project company.

concern to the project company and the lenders. This is where guarantees from host governments or IFIs, including the World Bank, become important.

Critical to the design of PPPs is the way risks are allocated between the partners in the PPP. In fact, effective risk allocation is key to success in any PPP. A general principle is that risk should fall on the party that is abler to do something about it. Risks in PPPs tend to be allocated on the basis of commercial and negotiating strength. The stronger party will allocate risk that it does not want to bear to the weaker party. Efficient allocation of risk will generally result in a more successful and profitable project and will benefit each of the parties involved (Delmon, 2009).

In order to minimize the market risk from the project company and the project lenders, an offtake purchase agreement, or in the case of a power project, a power purchase agreement, may be made. This is to create a secure payment stream which will be an important basis for financing the project. The offtake purchaser may also be the grantor, or a government entity such as a public utility, in which case the offtake purchase agreement and the concession agreement may be one and the same document (Delmon, 2009).

The lenders will want project risks to be allocated to project participants, i.e., the construction contractor and the operator but not the project company, which is their debtor. The project company will enter into a contract with the construction contractor in order to divest its obligations to the grantor to design, build, test, and commission the project. Completion risk for the project should be allocated to the construction contractor. In the case of a turnkey project, completion and performance risk should fall on the construction contractor.

If the main risks are associated with poor management of the service, shifting the risk to the operator could provide the right incentives to ensure that the project delivers. If risks are related to changes in policies, then the government should bear the risk. This is because the project company will not generally be able to manage political risk. The project company will ask the government to bear those risks, not necessarily to claim compensation at a future date but to pressure the government to avoid such risks and to minimize the probability that such risk events will occur.

Sharing the risks and the rewards in a fair and sustainable manner is important for the success of a PPP that may be operating for a few decades. During this time economic and financial crisis can hit as discussed in this book. This risk should be factored into the project. In turn, this can result in demands for higher returns for private sector participants. Crises in East Asia as well as in Latin America can teach valuable lessons to the private and public sectors, as well as to bilateral and multilateral providers of loans, equity and guarantees.

Reducing the risk of failure through efficient project company operations as well as good government policies is important for all parties involved. Shifting the risk to a weaker party is, however, questionable (e.g., if an insured private sector participant files a claim against the host government of a developing country without trying to resolve issues and then an insurer such as an IFI files a reclaim against the host government without making credible efforts to resolve the dispute).

PPPs place a strong demand on host government capacity to communicate and negotiate both with private participants and IFIs when these are involved. Low government capacity is an obstacle when using the PPP framework, but one must keep in mind that a weak government is also less likely to be able to run a project with 100 percent public ownership efficiently. When disputes occur between the private sector and host governments IFIs can, and should, play an important role in resolving disputes and help ensure the fair sharing of the risks and the rewards of the PPP for all the parties involved.

Chapter 8

Contested Multilateralism – The United States and China

The World Bank[63] was established at the Bretton Woods conference in New Hampshire in the US. When delegates from 44 nations assembled at Bretton Woods to create a new post-war global arrangement, a single country, the US, was the dominant power and held most of the world's gold reserves. Europe and Japan lay to a large extent in ruins and needed reconstruction. Because of its dominant position, the US was able to insist that the World Bank headquarters would be located in its capital, Washington DC. The Bank's headquarters ended up being located next door to the US Treasury Department. The World Bank president would also be appointed by the US president. [64] This arrangement would translate into a very strong influence of the US government over the World Bank, even beyond the power that comes from being the largest shareholder. Now virtually every country on earth is a member of the World Bank but the USA remains the most powerful member state.

Regional development banks, including the African Development Bank, the Asian Development Bank and the Inter-American Development Bank were later established to operate on a regional scale, often using instruments similar to those created at the World Bank. The European Bank for Reconstruction and

---

[63] Then called the International Bank for Reconstruction and Development (IBRD), reflecting the role it would play in reconstruction after World War II.

[64] Nothing in the World Bank's Articles of Agreement states that its president should be appointed by the president of the USA but this has become a tradition followed ever since its establishment.

Development (EBRD)[65] was created when the Berlin Wall fell and the Soviet Union collapsed.

The world has changed dramatically since Bretton Woods. For example, the World Bank Group now comprises five institutions [66] focusing on economic development on a global scale. Over the years, increasing pressures have emanated from developing and emerging countries to assume more power and influence within the World Bank Group. Since the 2008 crisis most economic growth in the world has taken place in emerging markets and new economic powerhouses have emerged. The G20 now challenges the authority of the G7. Joseph Stiglitz (2009) has for example talked about the power of the G20 as recognition of new realities "It's effectively recognition by the G7 that they don't have the money. The money is in Asia, the Middle East." (Stiglitz, 2009).

Most prominent among emerging market economics are the BRICS, i.e., Brazil, Russia, India, China and South Africa. These countries have taken a global lead by establishing the New Development Bank (the BRICS Bank). The Asian Infrastructure Investment Bank has also recently been established under Chinese leadership. The New Development Bank will be headquartering in Shanghai and the Asian Infrastructure Investment Bank in Beijing. Thus both institutions will be in China, the largest country among the new emerging powers.

As stated above, the New Development Bank's founding members are major emerging market countries: Brazil, China, India, Russia and South Africa. According to the founding countries of the New Development Bank, "The membership shall be open to members of the United Nations, in accordance with the provisions of the Articles of Agreement of the New Development Bank. It

---

[65] The EBRD was set up in haste to meet the challenge of an extraordinary moment in Europe's history: the collapse of communism in the Eastern bloc. In fact, a mere 18 months elapsed between the first mooting of the idea of a European bank in October 1989 and its opening for business with headquarters in London in April 1991.Urgency and the ability to respond to momentous events, whether it be the end of the Soviet Union, financial crises or the 'Arab Spring,' have been among the EBRD's hallmarks from the start. During the frenetic years of the early 1990s the EBRD's emphasis on the private sector as the main driver for change in Central and Eastern Europe was vindicated many times over. This was the period that established the EBRD's reputation as an expert on transition to the open market. It was involved in areas such as banking systems reform, the liberalization of prices, privatization (legalization and policy dialogue) and the creation of legal frameworks for property rights. See further http://www.ebrd.com/who-we-are/history-of-the-ebrd.html.

[66] The World Bank Group represents five institutions. These are: (i) the International Bank for Reconstruction and Development (IBRD), established in 1944, (ii) the International Development Association (IDA), established in 1960, (iii) the International Finance Corporation (IFC), established in 1956, (iv) the Multilateral Investment Guarantee Agency (MIGA), established in 1988, (v) the International Centre for Settlement of Investment Disputes (ICSID), established in 1966.

shall be open to borrowing and non-borrowing members" (New Development Bank, 2014). The New Development Bank will thus be a global institution.

Membership in the Asian Infrastructure Investment Bank will be open to members of the International Bank for Reconstruction and Development or the Asian Development Bank (Asian Infrastructure Investment Bank, 2015, p. 2). The Asian Infrastructure Investment Bank will thus also be a global institution in terms of membership, but with an Asian focus in its operations.

The stated purpose and functions of the New Development Bank is to "...mobilize resources for infrastructure and sustainable development projects in BRICS and other emerging economies and developing countries, complementing the existing efforts of multilateral and regional financial institutions for global growth and development" (New Development Bank, 2014). And "[t]o fulfill its purpose, the Bank shall support public or private projects through loans, guarantees, equity participation and other financial instruments. It shall also cooperate with international organizations and other financial entities, and provide technical assistance for projects to be supported by the Bank" (New Development Bank, 2014). The New Development Bank will thus have some of the instruments currently available at the World Bank as well as the African Development Bank, the Asian Development Bank, the European Bank for Reconstruction and Development and the Inter-American Development Bank. It is less clear what role it will have in engaging in policy dialogue with its beneficiary countries. It is supposed to complement existing institutions but it is hard to believe that it will not also compete and challenge the current global institutional structure.

Turning to the Asian Infrastructure Investment Bank, its Articles of Agreement state that: "The purpose of the Bank shall be to: (i) foster sustainable economic development, create wealth and improve infrastructure connectivity in Asia by investing in infrastructure and other productive sectors; and (ii) promote regional cooperation and partnership in addressing development challenges by working in close collaboration with other multilateral and bilateral development institutions" (Asian Infrastructure Investment Bank, 2015, p. 2).

And furthermore "[t]o implement its purpose, the Bank shall have the following functions: (i) to promote investment in the region of public and private capital for development purposes, in particular for development of infrastructure and other productive sectors; (ii) to utilize the resources at its disposal for financing such development in the region, including those projects and programs which will contribute most effectively to the harmonious economic growth of the region as a whole and having special regard to the needs of less developed members in the region; (iii) to encourage private investment in projects, enterprises and activities contributing to economic development in the region, in particular in infrastructure and other productive sectors, and to supplement private investment when private capital is not available on reasonable terms and conditions; and (iv) to undertake such other activities and provide such other services as may further these functions" (Asian Infrastructure Investment Bank, 2015, p. 2).

Finally "[t]he Bank may carry out its operations in any of the following ways: (i) by making, co-financing or participating in direct loans; (ii) by investment of funds in the equity capital of an institution or enterprise; (iii) by guaranteeing, whether as primary or secondary obligor, in whole or in part, loans for economic development; (iv) by deploying Special Funds resources in accordance with the agreements determining their use; (v) by providing technical assistance in accordance with Article 15; or (vi) through other types of financing as may be determined by the Board of Governors, by a Special Majority vote as provided in Article 28" (Asian Infrastructure Investment Bank, 2015, p. 8). Again the instruments are similar to those that the current IFIs have and with a strong focus on private sector cooperation in Asia.

The New Development Bank has the potential to become a global bank with global membership and global focus regarding projects in developing and emerging economies. The Asia Infrastructure Investment Bank, on the other hand, has a global membership but its operations will be regional, i.e., Asia. With both banks focusing on infrastructure and private sector cooperation they can potentially become important contributors to global efforts to scale up clean energy investments. The New Development Bank will potentially challenge the global leadership the US has established via the World Bank Group and the Asian Infrastructure Investment Bank will challenge the regional leadership that Japan has established via the Asian Development Bank.

THE NEW DEVELOPMENT BANK, THE ASIAN INFRASTRUCTURE INVESTMENT BANK AND CONTESTED MULTILATERALISM

Tensions between high income developed Western countries, on the one hand, and developing and emerging countries, on the other hand, have long existed within the World Bank Group, with the latter insisting on better management representation within the Bretton Woods institutions. Part of the reason for the emergence of these new potentially powerful institutions, the New Development Bank and the Asian Infrastructure Investment Bank, is the dissatisfaction of those countries with what they would consider a fair share in World Bank Group management. Zoellick (2009), then the president of the World Bank Group, for example argued that: "A strong G-2 within the G-20, and cutting across development lines, could form the cornerstone of a new multilateralism – a multilateralism that recognizes the realities of an international system born, not of nation-states alone, but of nation-states linked through economic interdependence. That modern multilateralism will require that rising economic powers have a larger say in how institutions such as the World Bank and the [International Monetary Fund] are run. This is both right and inevitable" (Zoellick, 2009 p. 10). But this did not happen, or at least it did not happen fast enough. It is notable that in 2012 the White House appointed Jim Yong Kim, an American physician and anthropologist, as president of the World Bank Group. Mr. Kim is of Asian origin but this was too little too late. The US does not seem to want to have its authority over the World Bank Group challenged.

Robert O. Keohane and Julia C. Morse (2014) have developed interesting theories on what they label as "contested multilateralism.[67]" According to them: "[c]ontested multilateralism involves the use of different multilateral institutions to challenge the rules, practices, or missions of existing multilateral institutions. More precisely, the phenomenon of contested multilateralism occurs when states and/or non-state actors either shift their focus from one existing institution to another or create an alternative multilateral institution to compete with existing ones" (Keohane and Morse, 2014, n.p.n).

In the case of the New Development Bank, major emerging powers have created an alternative global institutional headquarters in China to challenge the US lead and the Washington based World Bank Group. This may also result in the New Development Bank's founding nations shifting the focus from existing institutions, including with continued funding for the World Bank Group, most notably the IDA. The New Development Bank can also challenge regional IFIs.

In the case of the Asian Infrastructure Investment Bank the dominant emerging market countries, especially China, could shift the focus away from World Bank regional operations within the Asia Region. It could reduce its support for the Japanese led Asian Development Bank and reduce the role of other IFIs operating in Asia.

---

[67] I benefited from discussing this with Robert O. Keohane at the Woodrow Wilson School, Princeton University October 1 – 2, 2015.

It remains to be seen what financial and risk mitigation instruments these new institutions will use, if they engage in policy dialogue with their borrowing countries, and how much advisory and analytical work they will do. Time will also tell what their environmental and governance standards will be, but it seems likely that they, as new institutions, will be less bureaucratic and quicker to respond to their client countries than the World Bank Group and the Asian Development Bank, which suffer from layers of bureaucracy building up over the decades.

According to Keohane and Morse: "[s]tate power is a major determinant of whether coalitions have outside options. A group of dissatisfied actors that includes states with significant resources and institutional leverage will have an easier time identifying credible outside options than a coalition of weaker actors. If dissatisfied actors have an outside option, we should normally expect adaptation by the existing institution, since its authority and the scope of its impact will be adversely affected by the establishment of alternative organizations or practices" (Keohane and Morse, 2014, n.p.n.).

China, especially when joined by other major emerging countries, clearly has sufficient state power as well as the resources and institutional leverage to develop outside options, as they have now done in a dramatic way. Lawrence H. Summers, the former Secretary of the Treasury, commented as follows: "This past month may be remembered as the moment the United States lost its role as the underwriter of the global economic system……. But I can think of no event since Bretton Woods comparable to the combination of China's effort to establish a major new institution and the failure of the US to persuade dozens of its traditional allies, starting with Britain, to stay out of it."[68]

For some reasons the US within the World Bank and Japan within the Asian Development Bank did not make the necessary adaption to avoid the establishment of major competing institutions. Whatever adjustment they tried to make was too little and too late. Pressures on "traditional allies" did not work either. The result is the creation of alternative institutions with different practices without US membership.

---

[68] This past month may be remembered as the moment the United States lost its role as the underwriter of the global economic system. True, there have been any number of periods of frustration for the US before, and times when American behavior was hardly multilateralist, such as the 1971 Nixon shock, ending the convertibility of the dollar into gold. But I can think of no event since Bretton Woods comparable to the combination of China's effort to establish a major new institution and the failure of the US to persuade dozens of its traditional allies, starting with Britain, to stay out of it. See further at: http://larrysummers.com/2015/04/05/time-us-leadership-woke-up-to-new-economic-era/#sthash.1UTKrWpB.dpuf.

As Keohane and Morse state: "…existing institutions may fail to adapt. We posit that such failure to adapt occurs through one of two pathways. First, since dissatisfied actors have incentives to claim that they have outside options even when they do not, their incentives to misrepresent can generate problems of credibility. If the dissatisfied coalition is unable to make credible threats and promises about either its willingness or ability to leverage outside options to force change, it may take actions that result in contested multilateralism" (Keohane and Morse, 2014, n.p.n.).

It took many years for China to take the lead in cooperation with major emerging economies. However, the growing economic strength of emerging markets, most notable the BRICS, along with dissatisfaction with the current rich country dominance, triggered decisive action.

According to Keohane and Morse: "Even in situations where the dissatisfied coalition is credible, policy adjustment may not occur if there are diverging state interests, or ideational or institutional constraints. Veto players—states, organizational bureaucracies, or other crucial actors—may block changes desired by the challenging coalition because they see their interests—on substantive policy or with respect to institutional authority—threatened by change. Established conceptions of institutional roles and purpose are often persistent" (Keohane and Morse, 2014).

The USA was not willing to adjust even after decades of tensions and not able to control its allies, not even its closest allies at Bretton Woods, including the United Kingdom. Regardless of the tensions between the USA and the BRICS countries the establishment of the New Development Bank and the Asian Infrastructure Investment Bank could result in increased capital mobilization for clean energy needs in developing and emerging countries, especially in Asia, which has the largest unused geothermal resources in the world. That would be a welcome development.

Chapter 9

THE ICELANDIC GEOTHERMAL CLUSTER AND CROSS BORDER ENGAGEMENT

Iceland's progress in utilizing geothermal energy for space heating and electricity production has received international attention. Moreover, Iceland has become a significant player in the global geothermal energy market and is a leading country in Europe in geothermal utilization.

Only a few decades ago most of Iceland's energy needs came from fossil fuel: coal and oil. Most electricity and space heating is now based on clean energy. Al Gore described this transition in his book "Our Choice" as follows: "Iceland responded to the oil shocks of the 1970s by converting to domestic resources, virtually every building in the entire country is heated by the hot water resources close to the surface of the tectonically active land" (Gore, 2009, p. 109).

This transition to clean energy also meant that local capacity in geothermal energy utilization was developed. Several Icelandic companies and institutions now possess considerable experience in exploring geothermal sites, including drilling, as well as constructing and operating geothermal power plants. Additionally, service providers as well as research and education institutions have supported geothermal development in Iceland (see Table 12).

In November 2010 a conference in Reykjavík, attended by some 900 participants, discussed the potential of an Icelandic geothermal cluster to enhance Iceland's competitiveness and create a new engine of Icelandic economic growth. Among the participants was the leading global scholar on clusters, Professor Michael E. Porter from Harvard Business School, and the

President of Iceland, Dr. Ólafur Ragnar Grímsson[69], who has taken a strong interest in geothermal development and in Iceland's potential to become a global leader in this area.

Development of the cluster has continued and in February 2013 a legal entity was established for the Icelandic geothermal cluster initiative.[70] According to the website of Iceland Geothermal, the main role of the Iceland Geothermal Cluster Initiative is to promote Iceland "*As the land of geothermal energy and geothermal utilization*" (Iceland Geothermal, 2016). The purpose of Iceland Geothermal is to stimulate competitiveness within the Icelandic Geothermal Cluster, add value, and improve utilization of Iceland's geothermal energy.

Table 12. Some companies and institutions in Iceland with capacity in geothermal development. This list is only indicative and by no means exhaustive

Activity	Company/institution
GeoScience	ISOR, Mannvit, Vatnaskil
Technical Consulting	Mannvit, Verkís, Efla, Reykjavík Geothermal, Landsvirkjun Power, Nýsköpunarmiðstöð Íslands, Kemía
Business Consulting	KPMG, Capacent Corporate Finance, Íslandsbanki
Drilling	Jarðboranir, Ræktunarsamband Flóa og Skeiða
Construction	ISTAK, ÍAV, Loftorka, Eykt, Framtak, Héðinn, VHE, Altak, Rafal, RST Net
Energy Audit & Law Firms	KPMG, PwC, Deloitte, Lex (law firm), Logos (law firm)
Financing	Arion banki, Íslandsbanki, Landsbankinn
Geothermal Research	ISOR, Mannvit, Vatnaskil, Utilities, Universities
Research Funding	Orkusjóður, GEORG - Geothermal Research Group , Landsvirkjun Research Fund, Orkuveita Reykjavíkur Energy Fund, Rannís – Rannsóknamiðstöð Íslands
Training and Education	University of Akureyri, University of Iceland, Reykjavík University, Reykjavik Energy Graduate School of Sustainable Systems – REYST, Keilir – Atlantic Center of Excellence, Geothermal Training Programme - United Nations University (GTP- UNU)

Source: Constructed by the author

[69] Dr. Grímsson was president of Iceland from 1996 until leaving office on 1 August 2016.

[70] See further: http://www.icelandgeothermal.is/iceland-geothermal/history/.

The main goals of Iceland Geothermal "are to create new opportunities within the geothermal energy sector, facilitate cooperation with the aim of exporting services and building new partnerships, create a strong global geothermal value chain to enhance geothermal utilization worldwide. Provide benefits to the sector as well as to developing countries and protecting and valuing the environment". (Iceland Geothermal, 2016).

It is understandable that cross border engagement is being considered as an option for increased growth in the Icelandic geothermal sector. For example, a report prepared before Icelandic Geothermal was formally established entitled: *The Icelandic Geothermal Cluster; Cluster; Mapping and Mobilization,* states that "[f]oreign markets are important for Icelandic geothermal companies, especially in times when uncertainty and stagnation prevail for geothermal projects in Iceland" (Gekon, 2011, p. 35). This report also states that "[t]he economic collapse and a certain opposition to any further investment in geothermal projects in Iceland, however, have created an uncertainty about further development of Iceland's geothermal energy knowledge" (Gekon, 2011, p. 16). The report notes that "[i]t is easier to get a project if one can help with its funding. However, the difficulty of Icelandic parties with the financing of projects is one of the weakest aspects of the cluster. One way to address this weakness is to strengthen the cluster's cooperation with foreign investors and financial institutions in a systematic manner" (Gekon, 2011, p. 35). Finally, the report states that "So far Icelandic participation in international projects through ownership has not been successful. This is explained in part by the parties not having been sufficiently coordinated or critical of the projects that had been selected." The report does not explain what projects were examined to arrive at this conclusion.

For Iceland to engage successfully in global markets, certain issues need to be analyzed, issues such as coordination and formation of consortiums in emerging markets, as well as cooperation with foreign investors and financial institutions. Given that these are large, long term investments, funding and risk mitigation are also critical issues that need to be properly addressed and analyzed.

CLUSTERS - DEFINITION AND THEORETICAL CONSIDERATIONS

It is well known that economic clusters exist in virtually every industry and in every part of the world. But what exactly is a cluster? According to Professor

Porter, clusters are geographic concentrations of interconnected companies, specialized suppliers, service providers, firms in related industries, and associated institutions in a particular field that compete but also cooperate (Porter, 2000).[71] It seems reasonable to assume that society and industry could reap some benefits from reaching critical mass in experience and interactions in one place in a particular field.

Theoretically the assertion is that significant advantages accrue to companies from being in proximity to complementary products and services within reach of all the suppliers and partners in the product value chain. The emphasis on location and geographic concentrations, though, seems to contradict modern and global thought on the mobility of capital and knowledge. This seems like a paradox in an era of global competition. Here, however, competitive advantages are gained through interconnected companies and institutions locally, with competitiveness driven by the strength of the cluster, not merely the strength of individual companies. According to Czinkota et.al, cluster theory suggests that competition is altered in at least three ways when clusters form successfully: (i) by increasing the productivity of the companies based in the area; (ii) by driving and supporting the momentum of innovation in the area; and (iii) by stimulating creation of new companies and new configurations of business in the area (Czinkota, et al., 2009).

THE PRESIDENT OF ICELAND AND CROSS BORDER ENGAGEMENT IN DEVELOPING AND EMERGING MARKETS

During the geothermal conference in Reykjavík in November, 2010 the President of Iceland, Dr. Ólafur Ragnar Grímsson, made a memorable speech. It is worth quoting some of the statements he made to get a flavor of the ambitious visions expressed at the time. When talking about India, the president asked "Is it true that we can achieve enormous success in a relatively short time. I have talked to people in India for many years about geothermal energy," (Grímsson, 2010, p. 1). In his speech he also mentions China and Russia. "We have a window of opportunity for only the next five years or so. If we utilize it, there will be enormous opportunities, because it would take others years or decades to catch up while we keep running. But of course if we stop, they can

[71] For more discussion on clusters see for example, Porter, 1998, 2000, 2010 and 2016; Ketels and Memedovic, 2008; Ketels, 2010).

do the same thing as we are now doing, and perhaps beat us" (Grímsson, 2010, p. 3).

According to the president, many countries are waiting for cooperation with Iceland. As he says, "China, India, East Africa, Central America, Slovakia, Hungary and parts of the United States are now eagerly and actively looking to Iceland and asking themselves the question: How can we cultivate this co-operation?" (Grímsson, 2010, p. 5).

It is clear that the president was talking about cross border engagement with Icelandic involvement – an exporting cluster – but he did little to define exactly what this engagement would involve. Some attempt was made when he said: "We can define our partnerships in many ways. We could obtain a small shareholder stake in these products. We could build what I sometimes call elementary district heating systems in so many Chinese cities that it would be difficult to count them. If we obtain just a tiny percentage of that transformation in China, it would amount to a major economic input into the Icelandic economy" (Grímsson, 2010, p. 5). It is hard to fully understand what exactly this means but being a shareholder would normally require not only providing advice or selling technical expertise but also cross border capital investment.

PORTER AND THE GEOTHERMAL CLUSTER AS AN ENGINE OF RENEWED GROWTH

Michael E. Porter also made a comprehensive presentation during the geothermal conference in Reykjavík in November, 2010. While his comments were made shortly after the global economic and financial crisis hit Iceland (in October 2008) some obstacles still remain, including capital controls. A large follow up geothermal conference took place in Iceland in March 2013 but for some reason Porter did not attend this time.[72] This was unfortunate given that Porter had initiated much of the work in mapping and analyzing the Icelandic Geothermal cluster and its future potential. His presence in 2010 undoubtedly helped mobilize a large number of participants for this endeavor at the beginning. The discussion below is based on his 2010 presentation, which still remains highly relevant for the cluster.

When discussing investment to leverage Icelandic expertise in the geothermal sector, Porter stated that "lack of capital is a key constraint" (Porter,

[72] Michae. E. Porter did attend the Icelandic Geothermal Conference in Iceland in 2016 as will be discussed in a later section of this chapter, http://www.geothermalconference.is/

2010, p. 28). This was not surprising given that Iceland's major banks had all collapsed after the crisis and many companies in the geothermal sector were facing financial difficulties. When commenting on the potential for exporting services, Porter observed that Icelandic "[c]ompanies tend to lack size and capital to lead large projects" (Porter, 2010, p. 28). In addition, and related to his discussion on operational management, Porter correctly commented that "[s]kills (are) more technical than commercial" (Porter, 2010, p. 28).

These are all important observations. It is true that Icelandic companies tend to be small and even the biggest companies are faced with capital constraints, including large companies such as Landsvirkjun and Orkuveita Reykjavíkur. It is doubtful whether these companies could or should try to lead large cross border projects in emerging markets or developing countries. They have limited experience in this area except for Orkuveita Reykjavíkur, which, through Reykjavík Energy Invest, tried to engage in Djibouti and failed, wasting Icelandic taxpayers' money (see discussion below).

Besides, while Icelandic technical and engineering skills seem strong, the financial and commercial expertise necessary to mobilize resources for energy projects is limited. Moreover, keys to success include knowledge of, and skills in, applying risk mitigation instruments for capital mobilization in emerging markets, which often have a difficult investment and business climate.

Porter also emphasized the need to "[c]larify the role of publicly-owned companies in exports" (Porter, 2010, p. 32). This is an important point, especially for Landsvirkjun, which is in government ownership and Okruveita Reykjavíkur, owned by municipalities. Why should companies under public ownership take risks through cross border activities and pass the bill to the taxpayer?

Porter also talked about how important it was to "[i]dentify potential international partners" (Porter, 2010, p. 32), to "[a]ddress capital shortages" and of the "creation of a special financial instrument with government or foreign partners" (Porter, 2010, p. 32). This is especially important for companies coming from a small country that has capital controls, lacks funding and experience in emerging markets and can be vulnerable in dealing with host governments from emerging countries that are much larger.

In spite of these obstacles, Porter expressed strong confidence in the Icelandic Geothermal Cluster, including cross border engagement. He took as an example Huston, Texas, which has lost all its oil and gas but remains the global capital of oil and gas technology. As Porter noted, Huston is now exporting knowledge, skills, technology and expertise in capital and project management.

While Porter saw potential in growing domestic resources in Iceland, he saw a bigger long run opportunity for the Icelandic Geothermal Cluster in deploying skill, technology and expertise cross-border. According to Porter, someday soon Icelandic companies and Icelandic partnerships should be operating geothermal facilities all over the world. And as he stated, "We have every opportunity to be one of the globalizers of this business." (Porter, 2010). Porter was thus, like the president of Iceland, optimistic about potential in cross border engagement.

THE ICELANDIC GEOTHERMAL CLUSTER

Many Icelandic companies and institutions possess knowledge and experience in utilizing geothermal energy for space heating and electricity production. They could form an Icelandic geothermal export cluster where they would not only compete with each other but could also cooperate and potentially enhance each other's international competitiveness. Some of those companies and institutions are listed in Table 12 above.

There can be little doubt that these companies and institutions could also contribute to geothermal projects in developing and emerging countries. In fact, many of them are already active in some capacity in cross border operations and have been so for many years. The question is more how companies from small countries such as Iceland can contribute to the transition to clean energy in developing and emerging countries in a more comprehensive way and on a larger scale. These companies tend to be small and the economic and financial crisis in 2008 negatively affected the balance sheets of many Icelandic firms; indeed, some have not fully recovered. The largest companies, Landsvirkjun and Orkuveita Reykjavíkur, are in public ownership and the government needs to clarify what role, if any, they might play in geothermal development in developing and emerging countries. Icelandic companies also do not produce much machinery or equipment that could support geothermal development. Nevertheless, the geothermal know-how that now exists in Iceland and has been developed over several decades provides a strong platform for further engagement. Some staff members of geothermal companies have long and diverse experience plus proven technical skills. Based on this capacity a geothermal cluster is being developed that could engage cross border.[73] In

[73] Iceland Geothermal is a non-profit organization established in February 2013 around an industry-driven cluster cooperation in the field of geothermal energy, See further:

addition to space heating and electricity production, Icelandic companies have also made progress in multiple use activities such as fish farming, greenhouses, swimming pool heating, tourist lagoons, and so on. Some of these activities could be useful in developing countries; for example, drying fish is one area with high potential.

Participants in the Icelandic geothermal cluster have many challenges to overcome. Among the most difficult is the capital constraints they are faced with as well as proper risk mitigation in developing and emerging countries. They could potentially cooperate with the IFIs discussed in this book. However, while Iceland is a member of the World Bank Group and the European Bank for Reconstruction and Development, it is not a member of the African Development Bank, the Asian Development Bank and the Inter-American Development Bank. All the IFIs are large and carry out extensive feasibility studies before they move on with a project. They are bureaucratic, and project approvals take time. It is doubtful whether these long processing times fit well with the Icelandic mentality.[74] The Icelandic Ministry for Foreign Affairs conducts the relationship with the World Bank Group but the Ministry of Finance and Economic Affairs conducts the relationship with the European Bank for Reconstruction and Development (EBRD). Both ministries only have a small staff to cooperate with these large IFIs so it makes little sense to divide those responsibilities between two ministries. When writing this book, the author asked the Ministry of Finance and Economic Affairs for the rationale for placing World Bank Group and EBRD activities in two different ministries. No response was received.

---

https://docs.google.com/file/d/0By20q4BSBHqeTy1EWTU2ZWlpTmc/edit?usp=sharing&pli=1.

[74] The president of Iceland described this well when he was praising the Icelandic banks that collapsed shortly after his speech. When talking about the Icelandic approach the president said "On numerous occasions I have also emphasized how Icelandic society, including our history and traditions, has produced a modern business culture that has proven to be very favourable when meeting the competitive challenges of our times" (Grímsson, 2006, p. 3). And then the president goes on to describe some of those qualities of modern Icelandic business culture and talks about "The inclination to focus on results rather than a process: to go straight to the task and do the job in the shortest time possible" (Grímsson, 2006, p. 4). This would not fit well with the long preparation time often used by international financial institutions. The president goes on to say: "The absence of bureaucracy and our lack of tolerance for bureaucratic methods. Perhaps because there are so few of us, we have never really been able to afford extensive bureaucratic structures" (Grímsson, 2006, p. 4). This approach proved to be extraordinarily expensive for Iceland in the case of the banking sector, which eventually failed. If Icelandic investors want to gain the trust of international financial institutions and other reliable partners, this behavior must change. They need to learn.

Iceland did, via the Icelandic International Development Agency, cooperate on the Geothermal Exploration Project, with the Nordic Development Fund and the World Bank. The Icelandic International Development Agency was recently merged with the Ministry for Foreign Affairs. The Geothermal Exploration Project was jointly funded by the Icelandic International Development Agency and the Nordic Development Fund.[75] The stated objectives of the project are to assist countries in the East Africa Rift Valley in conducting surface exploration studies and build capacity and expertise in the field of geothermal development and utilization.

The project aims to assist countries to have: "(i) a realistic assessment of potential geothermal sites, (ii) a plan for further action where applicable with drilling targets identified, and (iii) capacity to move forward on the basis of those plans and submit exploration drilling projects into funding pipelines" (Icelandic International Development Agency, 2014, n.p.n.).

The project is the initial phase of the Geothermal Compact partnership, initiated jointly by the Ministry for Foreign Affairs in Iceland and the World Bank. The World Bank's Energy Sector Management Assistance Program serves as the focal point at the Bank for the Compact. The Icelandic International Development Agency (now the Ministry for Foreign Affairs) liaised with the World Bank on a country basis as well as through annual coordination meetings. Eleven countries have submitted requests for participation in the project, Ethiopia, Zambia, Tanzania, Burundi, Rwanda, Kenya, Djibouti, Uganda and Malawi, Comoros and Eritrea.

As noted above, the Agency has merged with the Ministry of Foreign Affairs and it is unknown how this will affect cooperation with the World Bank and the Nordic Development Fund. It is also notable that while Iceland emphasizes cooperation with African countries on geothermal development, it is not a member of the African Development Bank. Recently, however, the Icelandic government decided to become a member of the Asian Infrastructure Investment Bank. The Icelandic Ministry of Finance and Economic Affairs will conduct the relationship with the Asian Infrastructure Investment Bank and appears to have made the decision to join the bank without any feasibility study and without considering joining other regional development banks, including

[75] The Geothermal Exploration Project is jointly funded by The Icelandic International Development Agency (now merged with the Icelandic Ministry for Foreign Affairs) and the Nordic Development Fund. The main objective is to assist countries in the East Africa Rift Valley in conducting reconnaissance and surface exploration studies and building capacity and expertise in the field of geothermal development and utilization. See further: http://arsskyrslur.iceida.is/en/2014/regional-offices/geothermal/.

the African Development Bank, the Asian Development Bank and the Inter-American Development Bank.[76] Cooperating with African countries on geothermal development and joining an Asian investment bank is yet another example of the confusion, absence of coordination and lack of vision within the government of Iceland in terms of global participation in geothermal development. It would probably have been more realistic to support geothermal development either in Africa or in Asia.

CROSS BORDER GEOTHERMAL ENGAGEMENT. THE CASE OF REYKJAVÍK ENERGY INVEST IN DJIBOUTI

Over the past forty years, many attempts have been made to develop geothermal resources in Djibouti. Reykjavík Energy Invest's engagement is one example of a failed attempt as discussed in a recent project appraisal document from the World Bank (World Bank, 2013d).

According to the World Bank, the President of Iceland offered support to the President of Djibouti in 2007 to develop Djibouti's geothermal resources, an offer that was accepted. "The pledged support took the form of a proposal by Reykjavik Energy Invest … to take total project risk (including exploration risk) as an [Independent Power Producer] Developer and recoup investment and associated return through a 20-year electric tariff" (World Bank, 2013d, p. 70). This is highly unusual, for several reasons. First, the president of Iceland is a symbolic figure and does not have authority to make such commitments on behalf of the Icelandic government. Second, Reykjavik Energy Invest was fully owned by Reykjavík Energy, which in turn is owned by municipalities in Iceland, mainly Reykjavík, the capital city of Iceland. Third, it is most unusual, perhaps foolish, to take on the total project risks, including exploration, in Djibouti. In the event of failure, the cost would have to be borne by taxpayers in Reykjavík.

According to the World Bank, the parties signed legally binding Project and Power Purchase Agreements as well as an exclusive license for geothermal utilization in the Assal Rift (World Bank, 2013d). Before the deal was consummated, the 2008 global financial crisis hit Iceland and impacted Reykjavík Energy Invest's ability to fulfil its project obligations.

[76] The Ministry of Finance and Economic Affairs sent the author a one-page memo regarding the decision to join the Asian Infrastructure Investment Bank. This seems to be the only document prepared to make this important decision.

Reykjavík Energy Invest subsequently partnered with Contour Global and the IFC to try and amend the Project Agreement and Power Purchase Agreement and proposed a 13 cents/kilowatt hour tariff price. This was rejected by the Government of Djibouti. The proposal was rejected based on the observation that forty percent of the tariff reflected the cost of interest, which was significantly higher than originally planned (World Bank, 2013d). IFC, the private sector arm of the World Bank, thus got involved in the project.

Following this rejection, Reykjavík Energy Invest was reported, according to the World Bank, to be working with Sithe Global, an alternate Independent Power Producer, with the stated intention of proposing a lower tariff price structure. No alternate tariff pricing has been received by the Government of Djibouti from Reykjavík Energy Invest/Sithe Global since then (World Bank, 2013d). The development of Djibouti's geothermal potential thus once again came to a standstill.

According to the World Bank, the Government of Djibouti requested International Development Association (IDA) support and in October 2012 Reykjavík Energy Invest and the Government of Iceland agreed to provide all of the test data and studies previously performed by Reykjavík Energy Invest and its subcontractors while in pursuit of an Independent Power Producer based power plant in the Lake Assal region (World Bank, 2013d).

The information provided by Reykjavík Energy Invest will undoubtedly serve to reduce both the cost and the risk of the exploratory drilling project supported by the IDA. This, however, raises questions about the role and responsibility of the officials from Iceland who committed public funds for this failed project. The involvement of the two World Bank institutions, IFC and IDA, also raises questions. Why did the IFC not participate in the initial exploration costs with Reykjavík Energy Invest? Why did the World Bank not share the risks with Reykjavík Energy Invest? Why does the IDA then get involved in the project, potentially benefiting from Reykjavík Energy Invest's loss? The role and responsibility of Reykjavík Energy Invest and World Bank Group officials should have been investigated and those officials held accountable for the losses that in the end were borne by taxpayer customers of a public utility in Iceland.

The Djibouti experience shows a series of mistakes by officials in Iceland all the way from the very beginning. The initial commitment to take on total project risk (including exploration risk) as an Independent Power Producer Developer and recoup investment and associated return through a twenty-year electric tariff was foolish given the uncertainty that a viable project would ever materialize. The government of Djibouti and the World Bank could easily take advantage of this naïve offer. The Icelandic officials involved also appear to have taken advantage of the fact that any loss associated with this project could be passed on to Icelandic taxpayers and no questions asked. The Djibouti case is a good example of how Iceland should not engage cross border in geothermal development in developing and emerging markets.

Export Credit Agencies and Their Role in Risk Mitigation to Support Cross Border Trade Finance

Most developed countries have established export credit agencies to help finance export of their national goods and services as well as cross border investments. These agencies can provide guarantees in connection with projects involving deliveries of equipment and/or services to the project from the home country. This applies generally speaking to all export credit agencies supporting the export industry in their home country.

In Iceland such an agency exists and is called *Tryggingardeild útflutnings* (TRÚ). Export credit agencies can provide guarantees against both commercial and non-commercial risks in developing and emerging countries. These instruments can be quite suitable for supporting overseas investments in developing countries and emerging markets, including energy investments. TRÚ was to work in partnership with the Swedish export credit agency Exportkreditnamnden, which would assist the Icelandic agency to assess risks in host countries. Under Icelandic law TRÚ can provide guarantees and insurance up to 130 million SDR. This is a sizable amount of money. To keep the story short: TRÚ services have never been used by Icelandic exporters or cross border investors.

Developing and emerging countries are important for small states such as the Nordic countries for export and they all have export credit agencies to support their cross border trade. In contrast, Icelandic exporters and investors are not using the risk mitigation instruments that are available from TRÚ at all. In the case of Nam Theun 2, discussed in the next chapter, export credit agencies from Norway (Garanti-instituttet for eksportkreditt) and Sweden (Exportkreditnämnden), supported the project. Not having a functioning export credit agency is yet another institutional failure in Iceland that can make cross border investments in the energy sector problematic.

For more discussion about the role of export credit agencies in supporting cross border trade, including during times of crisis, as well as economic justifications for the existence of export credit agencies, see Dinh and Hilmarsson (2012a, 2012b and 2012c) and Hilmarsson and Dinh (2013).

The Icelandic Geothermal Conference, 2016

In April 2016 Professor Michael E. Porter participated in the Iceland Geothermal Conference.[77] The title of his presentation was "Unlocking the Potential of Geothermal Energy: Strategic Implications for Iceland, Other Nations and the International Community." As the title indicates, Porter's presentation focused both on the local agenda for Iceland as well as the global agenda.

Among issues on the action agenda for the Icelandic geothermal cluster according to Porter's recommendations was to develop financing mechanisms for geothermal projects abroad, including strong relationships with an international development institution (Porter, 2016). This is an important and relevant recommendation. Currently, though, Iceland is only a member of the World Bank Group, the European Bank for Reconstruction and Development and the newly established Asian Infrastructure Investment Bank.

It remains to be seen what the Asian Infrastructure Investment Bank will do but, as discussed earlier in this book, the World Bank Group and the European Bank for Reconstruction and Development have done little so far to support geothermal energy projects and spend about 94 percent on the least risky production phase. Clearly Iceland should cooperate more with those institutions, but they themselves need reform to partner with the private sector and they need to be more engaged. The signs are that geothermal development is receiving more attention within the World Bank and the European Bank for Reconstruction and Development.

On the Global Agenda, Professor Porter emphasized that financing structures must be better developed, reducing reliance on limited public sector funding sources (Porter, 2016). As has already been discussed in this book, developing a framework for public private partnerships is important for capital mobilization. PPPs in the geothermal sector, however, have a mixed record, as cases in Indonesia discussed in his book show, but it is possible to learn lessons from those cases, improve the incentive environment and avoid cases of moral

[77] Iceland Geothermal Conference 2016.

hazard. IFIs are well suited both to help with capital mobilization as well as with policy reform and they can also provide technical assistance in developing and emerging countries.

Chapter 10

CROSS BORDER CLEAN ENERGY INVESTMENT: FUNDING AND RISK MITIGATION - SOME CASES

This chapter discusses some cases of clean energy investment in developing and emerging market economies. This includes geothermal cases from Indonesia, Kenya and Turkey and hydropower cases from Lao and Uganda. These cases represent an effort to build new infrastructure to generate electric power without carbon dioxide emissions and is thus part of the global battle against climate change; for a cleaner environment and a healthier environment.

The Paris Agreement[78] signed in December 2015 commits "[m]aking finance flows consistent with a pathway towards low greenhouse gas emissions and climate-resilient development." Furthermore, it "[d]ecides that the Green Climate Fund and the Global Environment Facility, the entities entrusted with the operation of the Financial Mechanism of the Convention, as well as the Least Developed Countries Fund and the Special Climate Change Fund, administered by the Global Environment Facility, shall serve the Agreement." This means that from now on more funds should be available for clean and renewable investments especially in developing and emerging market economies.

This chapter thus starts by discussing these important international funds as the five international financial institutions (IFIs)[79] discussed in this book will be key implementing agencies for both the Green Climate Fund and the Global Environmental Facility. An increased flow of funds represents both an

---

[78] See the Paris Agreement at http://unfccc.int/resource/docs/2015/cop21/eng/l09r01.pdf.

[79] African Development Bank, Asian Development Bank, EBRD, Inter-American Development Bank and World Bank Group.

opportunity and a challenge for IFIs. This chapter also discusses Climate Investment Funds whose support is disseminated through the five IFIs.

Robert O. Keohane (2015) has remarked that "[t]he politics of climate change could be reframed if there were an emphasis on huge infrastructural projects that create jobs" (Keohane, 2015, p. 22). Of course large infrastructure projects cause pollution while under construction. However, these projects, which are typically large and long term, can also create multiple benefits over long periods.

Utilization of clean energy resources should certainly be part of the strategy for the battle against climate change and environmental degradation. In addition to important environmental benefits, clean energy investments generate jobs and profits important for developing and emerging countries to grow and eventually become high income and thus part of efforts to reduce poverty in the world. In addition, clean energy projects can result in multiple use activities, as discussed in the chapter on geothermal development in Iceland, that are both important for economic growth and can dramatically improve people's quality of life.

The project cases discussed in this chapter are important because they demonstrate how clean energy projects have been structured in various countries and continents, often under challenging circumstances. The cases also show innovative use of the funding and risk mitigation instruments discussed in this book.

THE GLOBAL ENVIRONMENT FACILITY[80]

The Global Environment Facility was established on the eve of the 1992 Rio Earth Summit, to help tackle the planet's most pressing environmental problems. Since then, the Global Environment Facility has provided US$14.5 billion in grants and mobilized US$75.4 billion in additional financing for almost 4,000 projects. The Global Environment Facility has become an international partnership of 183 countries, international institutions, civil society organizations, and the private sector to address global environmental issues. The Global Environment Facility along with the Green Climate Fund has been entrusted with operation of the Financial Mechanism under the Paris Agreement

[80]See further: https://www.thegef.org/gef/whatisgef.

signed in December 2015. The Global Environment Facility has eighteen implementing partners.[81]

GREEN CLIMATE FUND[82]

In deploying its resources, the Green Climate Fund will work through a wide range of institutions to finance projects and programs. To access funding, these institutions will go through a process of "accreditation," designed to assess whether they are capable of strong financial management and of safeguarding funded projects and programs against unforeseen environmental or social harm. The accreditation process is also designed to assess whether applicant entities have the ability to manage Green Climate Fund resources in line with the Fund's fiduciary standards for the scale and type of funding sought, as well as the ability to manage environmental and social risks that may arise at the project level.

Accredited entities carry out a range of activities that may include developing and submitting funding proposals for, and overseeing the management and implementation of, projects and programs; deploying a range of financial instruments within their respective capacities (grants, loans, equity, and guarantees); and mobilizing private sector capital. To date, twenty have been accredited to the Green Climate Fund including the IFIs mentioned in this book and the instruments they have for capital mobilization are highly relevant.

CLIMATE INVESTMENT FUNDS

The Climate Investment Funds are intended to be distinguished by a broad-based and inclusive governance structure.[83] A key feature of the Climate

[81] The Global Environment Facility's 18 implementing partners are: Asian Development Bank, African Development Bank, Development Bank of Latin America, Conservation International, Development Bank of Southern Africa, European Bank for Reconstruction and Development, Foreign Economic Cooperation Office - Ministry of Environmental Protection of China, Food and Agriculture Organization of the United Nations, Fundo Brasileiro para a Biodiversidade, Inter-American Development Bank, International Fund for Agricultural Development, International Union for Conservation of Nature, United Nations Development Programme, United Nations Environment Programme, United Nations Industrial Development Organization, West African Development Bank, World Bank Group, World Wildlife Fund US.

[82] See further: http://www.greenclimate.fund/ventures/accreditation/#get-accredited.

[83] The Strategic Climate Fund further designates Sub-Committees to govern each of its three-targeted programs: the Pilot Program for Climate Resilience, the Forest Investment Program,

Investment Funds' programming is the role of IFIs.[84] The largest development organizations in the world, IFIs offer grants and concessional financing as well as technical knowledge and experience in project design and implementation. The IFIs are to disburse Climate Investment Funds resources to support effective and flexible implementation of country-led programs and investment.

For the Climate Investment Funds, IFI involvement helps mitigate risk and encourages other actors to participate, thereby facilitating mobilization of co-financing and harmonization of policy support. The Climate Investment Funds leverage financing not only from the IFIs themselves, but also from other development actors (including UN agencies and bilateral development agencies) and the private sector, thereby demonstrating the potential of scaled-up, blended development financing. Climate Investment Funds-implementing IFIs include the regional IFIs; the African Development Bank, the Asian Development Bank, the European Bank for Reconstruction and Development, the Inter-American Development Bank and the global IFI, the World Bank Group (Climate Investment Funds, n.d.).

An IFI Committee for the Climate Investment Funds was established to facilitate collaboration, coordination, and information exchange among IFIs. The Administrative Unit facilitates the work of the Climate Investment Funds. It is housed in the World Bank Group's Washington, D.C. offices and comprises a small professional and administrative staff. The International Bank for Reconstruction and Development (IBRD) of the World Bank Group serves as the Trustee for the Climate Investment Funds. With the approval of the Trust Fund Committees, the Trustee makes commitments and transfers of Climate Investment Funds resources to the IFIs (Climate Investment Funds, n.d.).

The Clean Technology Fund,[85] with US$5.3 billion, is a funding window of the Climate Investment Funds. Its purpose is to promote transformation in middle income and developing countries by providing resources to scale up the demonstration, deployment, and transfer of low carbon technologies with a significant potential for long-term greenhouse gas emissions savings (Climate Investment Funds, n.d.). US$ 5.3 billion may sound a large commitment but on a global scale this is a small amount. In fact, the total resources of this fund amount to less than one US dollar per capita on the earth. Only a small fraction of those funds will be used for geothermal development. The Table 13 below shows commitments to geothermal development.

---

and the Scaling-Up Renewable Energy in Low Income Countries Program. Each Trust Fund Committee and Sub-Committee is composed of equal representation from contributor and recipient countries. The Climate Investment Funds Administrative Unit, IFI Committee, and Trustee support the work of both Trust Funds. See further, https://www-cif.climateinvestmentfunds.org/about/governance.

[84] On the website, IFIs are labeled as multilateral development banks. Both are correct.

[85] See further, https://www-cif.climateinvestmentfunds.org/fund/clean-technology-fund

Table 13. Geothermal Commitment of the Clean Technology Fund and the Scaling Up Renewable Energy Program[86]

Fund	Country	Project title	Sector	International Financial Institutions	Funding (US$M)	Co-financing (US$M)
Clean Technology Fund	Indonesia Philippines	Indonesia & Philippines: Private Sector Geothermal Program. Part of: Dedicated Private Sector Programs (DPSP II)	Private	Asian Development Bank	30.0	0.0
Clean Technology Fund	Indonesia	Geothermal Energy Upstream Development Project	Public	IBRD	49.6	2.810.5
Clean Technology Fund	Kenya	Concessional Finance Program for Geothermal Generation Part of: Dedicated Private Sector Programs (DPSP II)	Private	African Development Bank	30.0	0.0
Scaling Up Renewable Energy Program	Armenia	Geothermal Exploratory Drilling Project	Public	IBRD	8.9	108.6
Clean Technology Fund	Colombia	Utility-Scale Geothermal: Colombia: Financing and Risk Transfer Program for Geothermal Part of: Dedicated Private Sector Programs (DPSP I)	Public	Inter-American Development Bank	10.0	0.0
Clean Technology Fund	Chile	Technical Assistance for Sustainable Geothermal Development Project	Public	IBRD	3.0	0.5

[86] Under the Clean Technology Fund, the Dedicated Private Sector Programs were launched in 2013 to finance operations that can deliver scale and speed, while at the same time maintaining a strong link to country priorities and Clean Technology Fund program objectives. The Dedicated Private Sector Programs have utilized a programmatic approach where IFIs collaboratively identify private sector funding opportunities. Phase I of Dedicated Private Sector Programs was approved in October 2013 with an indicative allocation of US$150 million and phase II was approved in June 2014 with an indicative allocation of US$358 million, see further: https://www-cif.climateinvestmentfunds.org/dedicated-private-sector-programs.

Table 13. (Continued)

Fund	Country	Project title	Sector	International Financial Institutions	Funding (US$M)	Co-financing (US$M)
Clean Technology Fund	Caribbean Region	Utility Scale Renewable Energy: Geothermal - Sustainable Energy Facility (SEF) for the Eastern Caribbean. Part of: Dedicated Private Sector Programs (DPSP II)	Private	Inter-American Development Bank	20.0	0.0
Clean Technology Fund	Turkey	Turkey Geothermal Development Project Part of: Dedicated Private Sector Programs (DPSP I)	Public	IBRD	40.0	0.0
Clean Technology Fund	Turkey	Turkey: Geothermal Development Lending Facility Part of: Dedicated Private Sector Programs	Private	European Bank for Reconstruction and Development	25.0	0.0
Clean Technology Fund	Chile	Utility Scale RE-geothermal: Geothermal Risk Mitigation Program Part of: Dedicated Private Sector Programs (DPSP I)	Private	Inter- American Development Bank	20.0	0.0
Clean Technology Fund	Mexico	Utility Scale RE-geothermal-Geothermal Financing and Risk Transfer facility Part of: Dedicated Private Sector Programs (DPSP I)	Private	Inter-American Development Bank	20.0	0.0
Clean Technology Fund	Indonesia	Indonesia Geothermal Electricity Finance Program (IGEF)	Private	International Finance Corporation	50.0	1.720.0
Clean Technology Fund	Indonesia	Private Sector Geothermal Energy Program	Private	Asian Development Bank	150.0	2.475.0

Fund	Country	Project title	Sector	International Financial Institutions	Funding (US$M)	Co-financing (US$M)
Clean Technology Fund	Mexico	Geothermal Financing and Risk Transfer Facility	Public	Inter-American Development Bank	34.3	458.6
Clean Technology Fund	Chile	Geothermal Risk Mitigation Program (Financial Instrument Component)	Private	Inter-American Development Bank	30.0	264.5
Clean Technology Fund	Indonesia	Geothermal Clean Energy Investment Project	Public	IBRD	125.0	450.0
Scaling Up Renewable Energy Program	Ethiopia	Geothermal Sector Development Project (GSDP)	Public	IBRD	24.5	207.6
Scaling Up Renewable Energy Program	Kenya	Menengai Geothermal Development Project	Public	African Development Bank	25.0	360.0

Source: Climate Investment Funds , n.d.[87]

[87] See further, https://www-cif.climateinvestmentfunds.org/projects?field_related_country_target_id=All&field_mdb_tid=All&field_sector_tid=8&field_pp_sector_tid=All&field_related_fund_target_id=All&title

Indonesia Geothermal Development and Global Externalities

Indonesia has the largest estimated geothermal reserves in the world, and geothermal energy is an ideal option to diversify the country's power generation mix. Over 80 percent of current electricity generation in Indonesia is based on fossil fuels and increased utilization of geothermal energy could be an important step to hedge against the volatility of fossil-fuel prices. According to the World Bank less than four percent of Indonesia's geothermal resources has been developed to generate electricity[88] (World Bank, 2014, p. 3).[89]

Progress on geothermal development in Indonesia is of importance, not only for Indonesians themselves, but also for Asia and the world as a whole. This is because carbon dioxide emissions from one country affect all countries through their impact on global climate.

In Indonesia the World Bank cooperated with Pertamina Geothermal Energy.[90] This cooperation represents an effort to build new infrastructure to generate electric power without carbon dioxide emissions. The World Bank helped Pertamina Geothermal Energy kick-start the program by immediately developing fields where preparation is advanced and also helped strengthen the company's capacity to successfully implement its investments. Indonesia is poised to become the world leader in geothermal power generation capacity when the program is successfully implemented (World Bank, 2014, p. 3).

The project, supported by the World Bank, includes development of two geothermal fields: Ulubelu, located in the Lampung district in the southern part of the island of Sumatra, and Lahendong (Tompaso), located in the northern part of the island of Sulawesi. Pertamina Geothermal Energy plans expansions of approximately 110 megawatts in Ulubelu and 40 megawatts in Lahendong (Tompaso) (World Bank, 2014, p. 3).

The project is budgeted at a total cost of US$ 574.7 million. Approximately half the costs associated with development of the upstream steam field will be funded by Pertamina Geothermal Energy's own resources (equity) through the support of its parent company, Pertamina. The World Bank is providing loans

---

[88] Estimated geothermal power potential is 27,000 megawatts of which 1,189 megawatts had been developed by 2011 (World Bank, 2014).

[89] The World Bank Energy Sector Management Assistance Program and the Duke Center for International Development have prepared an excellent report entitled: Scaling-Up Renewable Geothermal Energy in Indonesia. The report was published by the World Bank.

[90] Pertamina Geothermal Energy (Indonesia) is a leading geothermal developer and a fully owned subsidiary of the state-owned oil and gas company, Pertamina.

totaling US$ 300 million. The financing package includes a US$ 175 million loan from the IBRD with a variable spread, loan[91] maturity of 24.5 years, and a grace period of nine years. The Clean Technology Fund, which is part of the global Climate Investment Funds, a group of funds established by the international community to promote climate-friendly and transformational investments, is providing a US$ 125 million loan on concessional terms, through the World Bank. The terms of the Clean Technology Fund loan are: 0.25 percent service charge (interest), total maturity of 40 years, and a grace period of 10 years (World Bank, 2014, p. 4).

According to the World Bank, the results of economic analysis of investment and operational costs suggest that the geothermal project is not competitive with an equivalent scale coal-based project when comparing the present values of the economic resource costs of investment and operations only. Based on the analysis, the present value of the investment and operational costs for geothermal is US$ 658 million, a full US$ 135 million more than the comparable coal-based option. However, these initial results do not take into consideration the environmental costs associated with utilizing coal or, conversely, the benefits of geothermal energy. Therefore, a complete economic analysis requires the inclusion of all externalities (World Bank, 2014, p. 5).

The problem that the government of Indonesia and the World Bank are faced with is that those who produce the emissions do not pay for that privilege, while those who are harmed are not compensated.

The emissions from coal based generation can adversely impact the health of people residing in the vicinity of power plants. This represents a negative local environmental externality that is costly. The appropriate response is to fix the market failure by repairing the flawed economic externality involved in climate change.

The development of geothermal energy also has global benefits, since it substantially reduces the emission of carbon dioxide when compared with a coal-fired power plant. The reduction in greenhouse gases, such as carbon dioxide, will positively impact global climate change. As the World Bank points out, it is important to note that this is a benefit that extends beyond the economy of Indonesia since the entire global community stands to benefit (World Bank, 2014, p. 6).

Due to the existence of a global externality (climate change benefits) and compensation provided by the international community through concessional financing, provided by the World Bank, the boundaries of the economic analysis

[91] LIBOR + variable spread.

were expanded to include the broader global community. The difference between the economic and the financial values represents externalities that accrue to a party other than Pertamina Geothermal Energy.

The economic analysis undertaken for this project confirms the rationale for undertaking the investment and explains why the international community needs to share the costs, given the project's global environmental benefits. However, without financial support, the project is not financially viable since the international and domestic policy environments do not "internalize" the "externalities" that would have adequately compensated the project's investor. This global market failure was addressed through a financing package that included a concessional loan from the Clean Technology Fund, which was instrumental in bridging the financial viability gap of the project (World Bank, 2014, p. 15).

The action taken in Indonesia and supported by the World Bank is especially important given the urgency to address climate change issues globally. Moreover, as the World Bank states, if urgent action is not taken, the International Energy Agency predicts that by 2017 the world will lock in the emissions of existing capital stock, leaving little room to maneuver and only far costlier options to consider thereafter for curtailing greenhouse gas emissions. This is well exemplified in Indonesia, where delays in expanding geothermal power generation will lead directly to development of coal-based power to meet baseload energy needs. Thus, this innovative investment, as well as other similar geothermal energy projects, represents global and local development imperatives (World Bank, 2014, p. 15).

Geothermal Development in Kenya: Olkaria III Geothermal

Kenya's economy is more diversified than most other countries in Sub-Saharan Africa. About 55 percent of Kenya's Gross Domestic Product comes from services, transport, finance, tourism, information and communications technology and trade sectors that are critically dependent on a reliable electricity supply.[92]

Only 25 percent of Kenya's population has access to electricity. The World Bank Group has been supporting the government's Least Cost Power

[92] See, for example; http://www.worldbank.org/projects/P122671/partial-risk-guarantees-ipps-kenya?lang=en.

Development Plan, which calls for an increase in the number of independent power producers and a more diversified energy mix in Kenya. The program benefits from a combination of guarantees from the Multilateral Investment Guarantee Agency, guarantees from the International Development Association and financing from the International Finance Corporation, all three World Bank Group institutions. According to the World Bank, these instruments play an important role in increasing investor confidence and in mobilizing the long-term financing needed to construct power plants (World Bank, 2015c).

The Kenyan government's plan called for a series of thermal and renewable independent power producers to replace the expensive, diesel-fired rental power plants currently in use. The first independent power producers in the program will be heavy-fuel oil plants, but subsequent independent power producers will use only low-carbon resources such as geothermal and wind, and the thermal plants will transition to peak-load operation. The challenge for the government of Kenya was attracting investors and lenders to deliver the program in the absence of sovereign guarantees, which was not possible under an agreed-on International Monetary Fund program (World Bank, 2015c).

One private company that has been particularly active in the development of geothermal projects in Kenya is Ormat Technologies, a US-based development company that established a subsidiary, OrPower4, in Kenya and through this entity has developed and currently operates the Olkaria III plant (Cambridge Economic Policy Associates, 2015).

The Olkaria III project is the first privately funded and developed geothermal project in Africa. It was enabled by a phased development strategy, and a combination of public and private financing and risk mitigation instruments that ensured the viability of the project (Climate Policy Initiative, 2015a). The project is located in the Olkaria geothermal field of the Rift Valley and is one of a series of geothermal developments. It is the only privately operated plant in the field[93] (World Bank, 2015c).

The Olkaria III project involved expansion of a base-load geothermal power plant. The plant came online with 8 megawatts in 2000. The plant has undergone several expansions, bringing current total generation capacity of the complex to 110 megawatts with the completion of phase three in February 2014. Electricity generated by the plant is sold under a power purchase agreement with the national power transmission and distribution utility—the Kenya Power & Lighting Company Limited. (World Bank, 2015c).

---

[93] Olkaria I, II, and IV are owned by Kenya Electricity Generating Company, a parastatal company 100 percent owned by the Government of Kenya (World Bank, 2015c).

Olkaria III cost US$ 445 million. Initially financed by equity in the late 1990s, the project was able to attract the debt needed for its expansion only in 2009 after renegotiation of the power purchase agreement and the attachment of a government security package to back the payments to the off-taker, the utility Kenya Power and Lighting Company (Climate Policy Initiative, 2015a).

According to the Climate Policy Initiative, the private developer Ormat provided equity financing with an initial US$40 million commitment in the years 1998-1999, which reached US$150 million in 2006. Ormat had to extend its equity commitment for longer than originally expected, securing debt financing only 11 years from the inception of the project. The current project finance structure relies heavily on debt from Development Finance Institutions, which now accounts for 85 percent of overall investment costs. Germany's Deutsche Investitions- und Entwicklungsgesellschaft mbH, together with KFW Development Bank, headed a financing consortium that refinanced Ormat's equity in Phase I with a US$ 105 million loan. The US Overseas Private Investment Corporation provided a 19-year tenor senior loan of US$310 million disbursed in three tranches used to finance Phase II and Phase III development and refinance part of the equity and debt provided earlier (Climate Policy Initiative, 2015a). The Multilateral Investment Guarantee Agency (MIGA) has supported the facility since 2000. The agency first provided a guarantee to Ormat Holding Corporation for its equity investment in OrPower 4, Inc., the project sponsor in 2000. In 2011, MIGA issued a guarantee of US$99 million to Ormat Holding Corporation for its equity investment in phase three. The guarantee is for a period of up to 15 years and covers the risks of transfer restriction, expropriation, and war and civil disturbance. This guarantee replaced the earlier MIGA guarantee covering investments in the first and second phases of the project (World Bank, 2015c).

The Least Cost Power Development Plan is expected to move Kenya away from historical reliance on hydropower energy for the bulk of its power generation, alleviating power shortages that have slowed economic growth in Kenya. According to the World Bank the government goal is to triple the national electricity supply of dependable energy to 3,000 megawatts by 2018, with emphasis on the development of alternative power sources—especially geothermal. This project is a step in that direction (World Bank, 2015c).

Geothermal Potential in Turkey

Turkey is a growing market for geothermal. In fact, in recent years installed capacity of geothermal power plants grew faster in Turkey than anywhere else in the world.

The European Bank for Reconstruction and Development (EBRD)[94] is becoming increasingly active in geothermal development. The EBRD was created in 1991 with its stated objective to promote transition to market economies in 34 countries from Central Europe to Central Asia. Since 2009, the EBRD has expanded its operations to include Turkey (2009) Egypt, Morocco, Tunisia, Jordan (2011) and Cyprus (2014). So far it has participated in financing eight geothermal power projects, seven of which are in Turkey.[95] Installed geothermal capacity in Turkey is 310 megawatts or 7 percent of the 4 gigawatt estimated potential. Western Turkey currently holds the greatest potential for development of geothermal resources, with Central and Eastern Anatolia largely unexplored (Herrera-Martínez, 2014).

Turkey clearly has strong potential in geothermal development but according to the EBRD suffers from technical, financial and legislative barriers (see Herrera-Martínez, 2014, p. 7). Among technical barriers are: limited expertise in project development and risk management; mid to low enthalpy resources; limited number of geothermal developers (infant market) despite strong potential. Financial barriers include: lack of financing for initial exploration and drilling phases; need for a strong balance sheet or pay high license premiums to overcome drilling risks; no VAT exemption guaranteed to the sector. Legislative barriers include: unclear regulatory framework; long administrative process prior to operation; difficult and heterogeneous licensing process; and difficulties for private sector involvement at initial levels of project development (e.g., surface exploration and drilling). EBRD's engagement in Turkey includes financing, technical assistance and policy dialogue.

According to the Climate Policy Initiative (2015) case study, the Turkey Gümüşköy Geothermal Power Plant is the first case where the private sector financed exploration of an unproven field in Turkey. The 13.2 megawatt project developed by BM Holding, a Turkish infrastructure company, was commissioned in 2013. The company demonstrated significant risk appetite in

---

[94] EBRD is owned by 65 countries and two inter-governmental institutions, with a capital base of €30 billion.

[95] Those Geothermal Power Plant projects are: Tuzla (2010), Gümüşköy (2012), Pamukören (2012), Babadere (2014), Germencik (2015), Alaşehir (2015) and Umurlu (2015). One project, Mutnovsky, is in the Kamchatka region (Herrera-Martínez, 2015).

undertaking early-stage exploration. BM Holding invested up to US$12 million (24% of total investment costs) in exploration and development prior to financial close, when debt financing of up to US$ 34.5 million (70% of the total costs) was secured from Yapikredi, a local commercial bank. Yapikredi sourced US$24.9 million of this debt from the Medium Size Sustainable Energy Finance Facility, an on-lending facility managed by the European Bank for Reconstruction and Development. The Government of Turkey's provision of a ten-year feed-in tariff ensured the project was financially viable (Climate Policy Initiative, 2015b).

Despite this growth, Turkey faces issues comparable to other countries seeking to develop geothermal capacity – specifically the ability of the private sector to take on the high risks associated with exploring and developing geothermal resources. Until 2013, 11 out of the 12 projects developed in Turkey were on sites where the government had already demonstrated that the resource was suitable for generating electricity and then put it out for tender. According to the Climate Policy Initiative, Turkey is now pushing for more private investment in the energy sector and the government has reduced drilling activity for geothermal exploration. More ambitious policy targets and a transition to a more private-sector led development model could help the sector realize its potential and would fit well with Turkey's current policy priorities (Climate Policy Initiative, 2015b). Like other IFI's, the EBRD is reluctant to fund geothermal exploration costs including initial drilling.

FUNDING HYDROPOWER PROJECTS IN AFRICA AND ASIA

Some cases indeed demonstrate that IFIs, bilateral Development Financial Institutions and export credit agencies can work with host governments and the private sector to mobilize funding for clean energy projects in difficult business and investment environments, where the private sector would generally hesitate to engage alone, and where local governments alone would have difficulty in mobilizing sufficient funding for large projects that require long term commitments. Among those cases are two hydropower projects in Asia and the other in Africa. One project is Nam Theun 2 (NT2) in Lao PDR and the other is the Bujagali project in Uganda.

Geothermal projects are different from hydropower projects, especially regarding initial development costs, but both have common characteristics in the form of large initial investment commitments with long repayment periods. The projects in Lao and Uganda, both developing countries, are located in challenging environments. The financial solutions for these projects demonstrate how funding and risk mitigation instruments from various multilateral and bilateral institutions can be successfully used to mobilize private sector funding for clean energy projects under challenging circumstances.

THE CASE OF THE NAM THEUN 2 HYDROPOWER PROJECT IN LAO PDR

Lao PDR is one of the poorest countries in South East Asia, with weak human capacity, governance, institutions and physical infrastructure. Nam Theun 2 is an example of how the public and private sectors can form a partnership and construct a major infrastructure project in the energy sector in a developing country with limited creditworthiness with support from IFIs and export credit agencies. Estimated project costs were US$1.25 million at financial close (excluding contingencies), equity 28 percent (US$350 million) and 72 percent debt (US$900 million).

The NT2 hydropower project was implemented by the Nam Theun 2 Power Company Limited (NTPC). The shareholders (equity holders) in NTPC were: Électricité de France International (35%), Italian-Thai Development Public Company Limited of Thailand (15%), Electricity Generating Public Company Limited of Thailand (25%) and Lao Holding State Enterprise (25%). Several IFIs provided loans to NTPC and/or guarantees to the private sector lenders: (i) multilateral institutions including the World Bank Group's IDA and Multilateral Investment Guarantee Agency (MIGA), (ii) bilateral agencies, and (iii) export credit agencies. Notably, a consortium of 16 commercial banks supported the project.[96]

A shareholders' agreement signed by Électricité de France International, the Government of Lao PDR, Electricity Generating Public Company Limited of Thailand, and Italian-Thai Development Public Company Limited sets out the rights and obligations of the shareholders, provides for the objective, establishment, management, and operation of the project company, NTPC, and agrees on the Articles of Association of NTPC. The shareholders' agreement

[96] The international commercial banks were: ANZ Bank, BNP Paribas, Bank of Tokyo Mitsubishi, Calyon, Fortis Bank, ING, KBC, SG and Standard Chartered.
The Thai commercial banks were: Bangkok Bank, Bank of Ayudhya, KASIKORNBANK, Krung Thai Bank, Siam City Bank, Siam Commercial Bank and Thai Military Bank.

lasts for 45 years from signing (World Bank, 2005). In the concession agreement, the Government of Lao PDR granted NTPC a concession to develop, own, finance, construct, and operate the hydroelectric plant and related facilities, and to transfer the project to the Government of Lao PDR at the end of the concession period, i.e., after 25 years (World Bank, 2005).

Nam Theun 2 is the largest ever foreign investment in Lao PDR and was the Asia Power Deal of the Year 2005. The project has an electric generating capacity of 1070 megawatts, of which 995 megawatts of the power was for export to Thailand and 75 megawatts for domestic use in Lao PDR. The power purchase agreements are between NTPC and the Electricity Generating Authority of Thailand, and between NTPC and Electricite du Laos.

A Head Construction Contract was signed between NTPC and Électricité de France International (the head contractor). This was a turnkey, price-capped engineering, procurement and construction contract (World Bank, 2005). The subcontractors were Italian-Thai Development Public Company Limited of Thailand, Nishmatsu Contracting Company of Japan, General Electric of the USA and Mitsubishi-Sumitomo Electric of Japan. The head contractor and the subcontractors are all reputable international companies.

IFIs played an instrumental role in enabling this project. In fact, the international dollar lenders to the project informed the NTPC that without political risk mitigation they would not be able to lend to the project. The Government of Lao PDR requested the World Bank Group to provide risk mitigation to support the international lending package (World Bank, 2005). IFI guarantees were thus key in lowering the project's risk profile sufficiently to attract the commercial financing needed.

Political risk guarantees were provided by MIGA (World Bank) and the Asian Development Bank. IDA (World Bank) also provided a partial risk guarantee. The NT2 partial risk guarantee is the first IDA guarantee to support hydropower development and is also the first project to use a mix of IDA, MIGA and Asian Development Bank guarantees. Debt guarantees were provided by IDA, MIGA and the Asian Development Bank supporting about US$126 million of private financing. Direct loans from IFIs were about US$144 million provided to NTPC (World Bank, 2005).

Loans were also provided by the Asian Development Bank, the European Investment Bank, the Nordic Investment Bank, Agence Française de Développement, Proparco and the Export-Import Bank of Thailand. The IDA and Agence Française de Développement also provided grants.

Nine International Commercial Banks and seven Thai commercial Banks helped fund the project. In addition, the NT2 project received export credit agency support from COFACE of France, Exportkreditnamnden of Sweden and its equivalent in Norway, Garanti-instituttet for eksportkreditt.

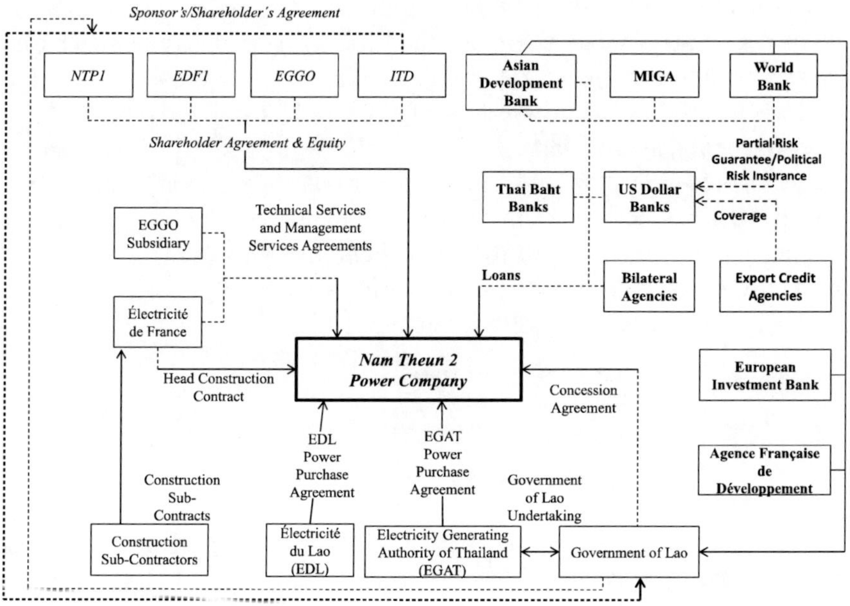

Source: World Bank, 2005

[i] The US$ senior debt facilities include those covered by political risk guarantees from IDA (World Bank), Asian Development Bank and MIGA; export credit agency support from COFACE of France, Exportkreditnämnden of Sweden and Garanti-instituttet for eksportkreditt of Norway, and direct loans from the Asian Development Bank, Nordic Investment Bank (NIB), Agence Française de Développement, PROPARCO and Export-Import Bank of Thailand. Nine international commercial banks (ANZ Bank, BNP Paribas, Bank of Tokyo Mitsubishi, Calyon, Fortis Bank, ING, KBC, SG and Standard Chartered) and seven Thai commercial banks (Bangkok Bank, Bank of Ayudhya, KASIKORNBANK, Krung Thai Bank, Siam City Bank, Siam Commercial Bank and Thai Military Bank) are providing long-term loans to NTPC.

[ii] Including EGAT as an off-taker; Engineering, Procurement and Construction contractors; project lenders and Government of Lao, as a shareholder in NTPC. Some risks may not affect all of the listed parties but only some.

i[ii] Excludes risks taken by GOL as an NTPC shareholder.

[iv] Natural force majeure: acts of God, earthquakes, fires, typhoons, and the like.

Figure 3. Nam Theun 2 Hydropower Project in Lao: Contractual Structure.

The Nam Theun 2 project can be viewed as a test case for infrastructure development in the developing world. It is an excellent demonstration of what is possible if the public and private sectors, supported by IFIs, team up and join forces. The use of IFI risk mitigation instruments is particularly interesting as it demonstrates how a modest commitment through such instruments can help mobilize much larger amounts of private funding.

The NT2 project was the world's largest private sector cross border power project financing, and the largest private sector hydropower project financing. Lessons learned from this landmark project would be a valuable study for all companies that intend to participate in infrastructure projects in developing and emerging market economies. In 2010 two senior managers from the World Bank published a book with comprehensive discussion about the lessons learned from Nam Theun 2 (see Porter and Shivakumar, 2010).

THE CASE OF THE BUJAGALI HYDROPOWER PROJECT IN UGANDA

Uganda is a developing country in Sub-Saharan Africa. Its first large scale independent power producer project, Bujagali Hydropower, was planned when the country was suffering from severe prevailing power shortages in the country slowing economic growth and reducing the wellbeing of its citizens. Bujagali is an example of how the public and private sectors can form a partnership and construct a major infrastructure project in the energy sector in a developing country with a challenging business and investment environment and with limited creditworthiness. In this case the project was built with support from IFIs and Development Financial Institutions. The total financing requirement for the project was US$798 million, of which US$ 627 million is financed by debt, and US$171 million financed by equity. The debt equity ratio is around 78:22 (World Bank, 2007).

The Bujagali 250 megawatt Hydropower project was implemented by Bujagali Energy Limited. The Implementation Agreement between the Government of Uganda/Uganda Electricity Transmission Company Limited sets out the terms on which the Government grants to Bujagali Energy Limited the concession to design, finance, construct, own, operate, and maintain the project (World Bank, 2007). The sponsors of Bujagali Energy Limited are Industrial Promotion Services (Kenya) Limited and SG Bujagali Holdings Ltd, an affiliate of Sithe Global Power, LLC (USA). The sponsors provided US$151 million equity and the Government of Kenya US$20 million.

Several IFIs provided loans to Bujagali Energy Limited and/or guarantees to the private sector lenders: (i) multilateral institutions including the World Bank Group's IDA, IFC and MIGA and development finance institutions. Two commercial banks, Standard Chartered and Absa, supported the project.

The project was developed on an independent power producer basis. It is developed, built, owned, and operated by Bujagali Energy Limited. The project sells electricity to Uganda Electricity Transmission Company Limited under a 30-year power purchase agreement (World Bank, 2007).

An engineering, procurement and construction contract was signed between Bujagali Energy Limited and Salini SPA. The proposed project was built pursuant to a fixed price, date certain, turnkey engineering, procurement and construction contract. The engineering, procurement and construction contractor, Salini SPA (Italy) (with Alstom Power Hydraulique (France) as a key subcontractor) was selected pursuant to a competitive engineering, procurement and construction contract selection process in accordance with European Investment Bank procurement rules (World Bank, 2007).

IFIs played an instrumental role in enabling this project. After unsuccessful attempts to develop the project in the late 90s, the Government of Uganda initiated a new bidding process, with the support of the World Bank, seeking a new project sponsor to develop the Bujagali project (World Bank, 2007). The IDA guarantee reduced the perceived risk in the project to such an extent as to allow commercial debt to be mobilized.

The debt facility consisted of a commercial loan of US$115 million, from the Standard Chartered and Absa banks, covered by a World Bank partial risk guarantee. The rest of the financing came from other multilaterals, such as International Finance Corporation, which committed US$130 million in loans, the European Investment Bank lent US$140 million, and the African Development Bank US$110 million. European development finance institutions financing consists of French development agency Proparco, with a US$73m loan, Deutsche Investitions- und Entwicklungsgesellschaft mbH/KfW of Germany with US$45m, and Dutch financier FMO with US$73m (World Bank, 2007). The Multilateral Investment Guarantee Agency (MIGA) provided an equity investment guarantee of up to US$115m for a 20 year period.

The Bujagali project is one of the largest private sector financed projects in Sub-Saharan Africa so far and the first of its kind in Uganda. It was awarded the "The Africa Power Deal of the Year 2007" by Euromoney "Project Finance Magazine."

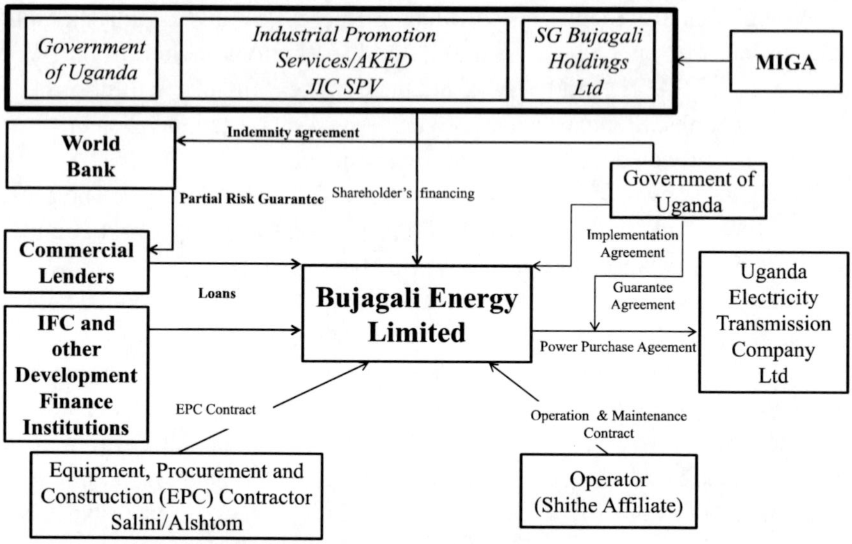

Source: World Bank, 2007.

Figure 4. Bujagali Hydropower Project in Uganda: Contractual Structure.

Along with the Nam Theun 2 project, also discussed in this book, it demonstrates what is possible if the government, the private sector and the international community use available multilateral and bilateral institutions to mobilize funding for clean energy in developing countries. The IDA guarantee reduces the risk faced by the commercial lenders. MIGA reduces the risk for the private equity providers. A loan from the International Finance Corporation further reduces the risk profile for this project. The other IFIs, the African Development Bank and the European Investment Bank and the development finance institutions also play a key role to make this project bankable.

Chapter 11

A TALE OF TWO CRISES AND THE URGENCY FOR ACTION

The Bretton Woods institutions, namely the International Monetary Fund and the World Bank,[97] established by the US and is allies in the aftermath of World War II, are still important global players. The role of the World Bank shifted from World War II reconstruction to development in developing and emerging countries. Currently, the World Bank is the largest and most important development organization in the world and has in fact been divided into five institutions commonly known as the World Bank Group. The International Monetary Fund still has an important global role to play, especially during financial crises and, unlike the World Bank, it can also assist developed high income countries. The World Bank's tasks as a development institution are typically long term compared to the challenges faced by the International Monetary Fund, which tend to be short term. Thus the International Monetary Fund is sometimes described as a firefighter.

To stay relevant, these two institutions need to adjust with time and respond to changing world needs. Two global challenges have received much attention in recent years. One is the global economic and financial crisis that hit in the fall of 2008 that also became a human crisis in many countries. This crisis still haunts countries and regions. The other crisis is the climate crisis that a growing number of scientists see as a threat to mankind and requires immediate and urgent global action.

It would be hard to argue that the International Monetary Fund's response to the 2008 global crisis was particularly slow or ineffective although one can

---

[97] Initially called the International Bank for Reconstruction and Development.

always find some who feel that more should have been done and faster. The International Monetary Fund acted as a firefighter. In contrast, the World Bank's response to the climate change crisis has been slow. The World Bank's clean energy project portfolio is small in a global context and many World Bank staff are frustrated with its slow response and the heavy bureaucracy. So far, the World Bank has done little in the battle against climate change. Other international organizations such as regional IFIs and the UN have done even less.

In a recent book entitled *A Global Life*, former World Bank president James D. Wolfensohn discusses this different response time of the two Bretton Woods institutions. In the book, Wolfensohn remarks that "[i]n contrast to our [World Bank] practice, the International Monetary Fund had a track record of acting quickly and decisively when it was needed. And to be fair, in any comparison of the two Bretton Woods institutions, this advantage had been built into the [International Monetary Fund] – its ability to respond quickly to financial crises was an essential part of its mission" (Wolfensohn, 2010, p. 310).

To some extent this is understandable, given the history of those two institutions, but to stay relevant institutions need to change when the world changes and needs different services from its global institutions. As time has passed the World Bank has had the tendency to become ever more slow and bureaucratic with the introduction of endless safeguards. As Wolfensohn says, "Because the loan approval process was so rigorous, with careful checks and balances built into our procedures, we rarely were able to move with immediate impact, even under streamlined procedures" (Wolfensohn, 2010, p. 310) and further "It was deemed to be slow, clumsy and maddeningly inept at addressing poverty and the environment. Some suggested that management appeared more concerned with preserving a top rating for its bonds than with the quality and effectiveness of the loans it was making" (Wolfensohn, 2010, p. 258). This is not a new concern. Gavin and Rodrik, for example, noted twenty years ago that "...the need to maintain the [triple-A credit] rating continues to act as a constraint on the nature of ideas about development that the [World] Bank can generate and nurture" (Gavin and Rodrik, 1995, p. 333). The World Bank has also become very risk averse. This is, for example, reflected in its small clean energy project portfolio.

Certainly, maintaining a good credit rating is an issue that the World Bank must consider in its actions. It must keep a healthy balance sheet and be prudent financially. However, if this means that it cannot respond to serious crises like climate change and make a significant contribution to a transition to clean energy, its operations become increasingly irrelevant to the most pressing needs of the world. One reason for this small clean energy portfolio is that the World Bank has been hesitant to use its guarantee powers to mobilize funds. One can wonder if the World Bank has become self-serving and too risk averse. The World Bank's main objective should not be to maintain itself and at best pretend to respond to pressing global issues. If it cannot respond, then other institutions are needed.

In addition, an ongoing struggle is in progress between the World Bank and its rich donor countries, especially the US. When discussing the urgency to respond to climate change, Nordhaus (2013) comments that "[t]he need to introduce effective policies has been particularly difficult in the United States, where ideological opposition has hardened even as the scientific concerns have become increasingly grave" (Nordhaus, 2013, p. 9). And as Wolfensohn states: "The struggle between the Bank and the US Treasury traced back almost to the [World] Bank's inception" (Wolfensohn, 2010, p. 322). At Bretton Woods the US was a dominant power. While some of its allies, including the United Kingdom, objected, the US felt that the World Bank's headquarters should be in Washington DC and that its government would to a large degree control what the World Bank did, when and how. The US Treasury's proximity to the World Bank headquarters brings power. Wolfensohn describes this well: "The US executive director of the [World] Bank, unlike other directors whose capitals were far away, could walk down the street as needed to attend top-level Treasury meetings, then return and exert a very strong influence on the [World] Bank president and in the boardroom." (Wolfensohn, 2010, p. 322). Today the US is not as dominant a power as it was immediately after World War II. Nevertheless, it is the most powerful country in the world and international cooperation does not work without US participation.

Wolfensohn also tells a compelling story in his book *Global Life*. One day when serving as president, Deputy Secretary Lawrence Summers at the US Treasury called him because of a Russian financial crisis demanding that the World Bank immediately commit US$6 billion to Russia's aid. Executive Directors from wealthy countries were instructed by their capitals to vote in favor. Wolfensohn was faced with a done deal. The resolution passed (Wolfensohn, 2010, p. 320-321). This shows how powerful the US is and how decisively it, and its allies can act if they consider immediate action is important. The US$6 billion approved by the World Bank board is twice the amount the World Bank, the African Development Bank, the Asian Development Bank, the European Investment Bank and the Inter-American Development Bank combined have committed to geothermal projects over a period of three decades. This shows that the World Bank can move quickly and decisively if its rich owners want it to move. Being slow, bureaucratic and clumsy is not unchangeable natural law.

Since Bretton Woods new powers have emerged, most notably China, whose economy may soon become larger than that of the US. The Chinese have learned that proximity brings power. They have taken a lead in establishing two new development banks – the Asian Infrastructure Investment Bank, with a regional focus, and the New Development Bank (BRICS Bank), with a global focus. One is located in China's capital, Beijing, and the other in Shanghai as discussed earlier in this book. This seriously challenges the global authority of the current IFIs, especially the US dominated World Bank and the Japan dominated Asian Development Bank.

The Urgency to Take Action

The global economic and financial crisis that hit in the fall of 2008 was a complete surprise for virtually every economic forecaster. Among the most important lessons from the crisis is that no one understood how fragile the system was.[98] The world was not prepared to respond to this catastrophic event. Rudi Dornbusch once said that financial crises take much longer to arrive than you think, and then they happen much faster than you could imagine. The same can apply to a climate crisis. Climate change can lead to a sudden irreversible crisis with severe consequences that are difficult to respond to.

No panacea is available to solve this problem, but it is important to take it seriously and take steps to remedy it. This includes investment in the clean energy sources available to us. International financial institutions (IFIs) are uniquely suited to deal with this problem as they can provide technical assistance, engage in policy dialogue with governments in developing and emerging market economies, in addition to making funding and risk mitigation instruments available to assist with capital mobilization.

But they are not taking urgent action. They are risk averse and appear to be more concerned about their own balance sheet and self-interest than addressing the threat mankind is facing. It would be simplistic to blame only their staff for this status. Their owners, especially the rich nations, run the show and decide what priorities IFIs have.

[98] Some economists and financial analysts claim to have predicted the crisis. What many of those have in common is that they predict crises all the time.

Being slow and bureaucratic are manmade obstacles. At Bretton Woods the International Monetary Fund was designed in such a way that it could respond to crises. During the 2008 economic and financial crisis it responded quickly. The World Bank can also respond quickly if the rich member states that hold most of its shares want it to do so and US support is key to results.

IFIs also need to take a broader view when considering action in the battle against climate change. Utilizing clean energy sources is not only part of the benefit. In addition to a cleaner and healthier environment, utilizing clean energy sources can have multiple benefits such as in the case of geothermal in Iceland and generate new value added sectors that can also be important for developing and emerging countries. In fact, adding value this way is how low income countries move on to become middle income countries and eventually high income countries, as Iceland did.

In his book *The Climate Casino*, Nordhaus (2013) states that "[c]limate change is no longer just geophysics and ecology; it has become economics and politics" (Nordhaus, 2013, p. 66). This is true and it is increasingly important that the response to the climate crisis be multidisciplinary, including not only natural scientists, but also financial analysts, economists, political scientists and other related disciplines. Currently, a multidisciplinary approach is lacking, as is abundantly evident, for example, in the geothermal sector if one attends international conferences on the subject. Those conferences tend to be dominated by natural scientists.

To increase participation in international climate actions Nordhaus has argued that "[a]n effective global arrangement will need an effective mechanism to encourage participation and discourage free riding. The most promising approach is to impose import tariffs on the products and services of nonparticipants" (Nordhaus, 2013, p. 8).

Imposing taxes and tariffs can certainly be part of the solution but it is also important to take multiple actions simultaneously. This includes providing incentives for promoting participation in clean energy investment that also create jobs and profits. As Keohane (2015) has stated, "…plausible political efforts to finance large infrastructure investments would generate self-interested incentives by the relevant construction industries to support vigorous action (Keohane, 2015, p. 22)." Furthermore, he maintains that "[t]he politics of climate change could be reframed if there were an emphasis on huge infrastructural projects that create jobs" (Keohane, 2015, p. 22).

Taxes and tariffs are unpopular and are likely to be met with resistance. Inevitably, huge investments in infrastructure will take place globally in the coming decades. Some of these will be supported by the IFIs including the new banks, the Asian Infrastructure Investment Bank and New Development Bank. It is of crucial importance that this infrastructure is climate friendly and that priority is given to use clean energy sources now available on earth. All the IFIs need to contribute to this effort.

Oil prices have dropped by over 60 percent since June 2014. According to Arezki and Obstfeld (2015) fossil fuel prices are likely to stay low for a long time. Low fossil fuel prices can discourage further innovation in and adoption of cleaner energy technologies and the result would be higher emissions of carbon dioxide and other greenhouse gases. It is important that policymakers do not allow low energy prices to derail the clean energy transition (Arezki and Obstfeld, 2015). Earlier transitions, from wood/biomass to coal and from coal to petroleum took a long time. We cannot afford such a long transition time today.

Overview of Some Sources of Funding to Promote Cross Border Geothermal Energy Projects

This section provides an overview of some possible sources of funding for cross border geothermal projects at an early, middle and late stage of project development. As discussed earlier in this book, high upfront development costs are associated with geothermal energy projects. At the initial phase risks are high but at the middle stage risks are medium and at the late stage the risks are low and projects may become bankable. Table 14 below provides an indication as to what funding sources and risk mitigation instruments might be feasible at each of the three stages of geothermal project development. Exploration and initial drilling costs are especially difficult for developing countries with small fiscal budgets and a weak tax base. Unfortunately, as shown in Table 10, IFIs have been avoiding this high risk in the carly phase of geothermal development. Here they could and should do much more in the shape, for example, of offering grants and concessional loans. Other donors should also become involved with grants or subsidized loans. During the middle and especially the late stage of project development, IFIs should make more use of their guarantee and insurance instruments for capital mobilization, for example attracting loans from international investment banks. Guarantees and insurance instruments could also attract equity contributions from international investors and investment funds.

Table 14. Overview of potential sources of funding for cross border geothermal projects at different stages of project development

Stage of project development	***Early***	***Middle***	***Late***
Activities	Exploration Initial drilling	Resource confirmation Field development Production drilling	Power plant Construction Commissioning
Risk of failure	*High risk*	*Medium risk*	*Low risk*
Potential funding sources	Balance sheet financing from sponsors Private equity from shareholders/investors Government cost sharing, concessional loans, grants, loan guarantees Donor grants, IFIs and bilateral donors Concessional loans/funds from international donors, including IFIs* (e.g., World Bank IDA terms)	Balance sheet financing or corporate bonds from sponsors Public equity Construction debt Loan guarantee by government Long term loan from IFIs* Guarantees from IFIs against non-commercial risk to facilitate commercial lending Export credit agencies **/trade finance. Guarantees and insurance against commercial and non-commercial risks. Buyer or Supplier credits. Direct loans	Long term loans from commercial sources/investment banks Long term loan from IFIs* Guarantees from IFIs to attract commercial loans (larger loan allocation, longer term, lower interest rate) Export credit agencies **/trade finance. Guarantees and insurance against commercial and non-commercial risks. Buyer or Supplier credits. Direct loans

*IFIs: International Financial Institutions: Lending, equity, guarantees, grants and policy reform
**Export Credit Agencies: Guarantees for trade finance and loans
Source: Constructed by the author

IFIs should be able to price guarantee and insurance instruments lower than the private sector because they often engage with governments in developing countries on multiple projects and programs, which gives them as multilateral institutions leverage *vis-à-vis* developing country governments. An example of additional risk mitigation is to allow host governments to own a stake in their own country's geothermal power plant, which gives them a stronger incentive to help ensure project success. They would thus become shareholders/equity holders in the project and good performance would result in dividend payments to the host government.

IFIs have so far made little use of their guarantee powers to mobilize funding for clean energy projects. High costs and bureaucracy are cited as major reasons, not only by outside observers, but also by staff and managers within the World Bank. Export credit agencies can also play an important role in facilitating trade in goods and services from providers in developed countries to geothermal projects in developing and emerging countries. Table 14 below provides an overview of some potential sources of funding for cross border geothermal projects.

Conclusion

International Financial Institutions as Key Instruments for Successful Global Climate Solutions

The battle against climate change requires international cooperation and coordination. Strong incentives exist for individual countries to free ride and the absence of a global government, or a strong international organization to enforce cuts in emissions, make this battle a sharply challenging problem to manage. As discussed in this book, rich countries have preferred to sign bilateral investment treaties with developing and emerging countries rather than having a global investment regime supported by an international organization. This makes long term clean energy investments in developing and emerging markets more problematic. International Financial Institutions (IFIs) tend to serve investors from rich countries rather than the interests of developing host countries. Self-interest and short-sighted (in)action on the part of rich countries make climate change a very difficult problem to manage. Weak institutions and governance problems in developing and emerging countries only add to this problem.

Part of the strategy in the battle against climate change ought to be transition to clean energy sources. This requires infrastructure investment to utilize these sources on a larger scale. This is currently not happening fast enough and investment in developing and emerging markets is often challenging because of underdeveloped business and investment environments. Emerging and developing countries have the largest underutilized clean energy sources in the world and they also have the strongest demand given both current economic and population growth. Clean energy investments like those discussed in this book, i.e.: geothermal and hydropower, are large, capital intensive, and long term.

Those projects often need sponsorship and funding from public, private and donor sources. This requires international cooperation and coordination, cooperation between the public and the private sector, as well as a fair sharing of risks and rewards.

The success of the World Trade Organization in preserving the long term benefits of an open global trading system is notable. One can argue that the General Agreement on Tariffs and Trade that later became the World Trade Organization is the most successful international organization ever established. In contrast, the current investment regime is mostly bilateral, with decentralized decision making, and no international organization supports the regime: no World Trade Organization is available for cross border investments.

In the absence of a global investment agreement, like the General Agreement on Tariffs and Trade and subsequently the World Trade Organization on trade, investors have turned to piecemeal solutions when protecting their rights in risky countries. This failure to create a global framework for investment is especially serious if one considers clean energy projects, which tend to be large, capital intensive and long term, including in the cases of geothermal and hydropower projects. The current investment regime seems mainly to be concerned with protecting investors from rich countries against government action in poor countries.

The failure to create a global investment regime traces all the way back to the Bretton Woods conference. In fact, one can hardly talk about any institutionalized cooperation on cross border investments on a global scale. Some regional agreements and interregional agreements address cross border investment but it is unclear yet if those will eventually result in a global arrangement supported by an international organization. Rich countries may be better off with the current arrangement, i.e., the interests of capital exporting countries can continue to dominate developing and emerging countries.

The International Bank for Reconstruction and Development, now the World Bank, was created at Bretton Woods. Its purpose was to support reconstruction in Europe after World War II but now its focus is on development in emerging and developing countries. It was intended to use guarantees to generate private sector investment but it never made much use of its guarantee powers. Later, regional development banks were also created with similar instruments, to a large extent introducing similar instruments to the World Bank but applying them regionally.

IFIs can be important partners not only with direct funding, i.e., loans and equity investment, but also increasingly through risk mitigation instruments. However, the insurance and guarantee instruments offered by IFIs to promote cross border investments are piecemeal solutions that could nevertheless be useful to leverage funding for clean energy investments. While they have not been used much to support cross border clean energy projects, some signs suggest that e.g., the World Bank may be stepping up its efforts. IFIs can also make important contributions via policy dialogue and technical assistance.

Increased investment in clean energy projects can only be part of the solution in the battle against climate change. Nevertheless, it is an important part of the strategy. Investment in infrastructure creates jobs and profits and can thus provide the right incentives for global action. Any global climate agreement needs to construct a finance regime that supports low-carbon development in developing and emerging countries. The climate finance regime should incentivize decarbonization.

The international community needs to ask two questions: How can we make the global investment regime more effective, especially in promoting clean energy investment in developing and emerging markets? How can we reform the IFIs to make them more effective in responding to the climate crisis?

Currently, new IFIs are being created under the leadership of large emerging market economies, namely the BRICS and within that group most notably China. Those new banks, the Asian Infrastructure Investment Bank and the New Development Bank have the potential to take decisive action in clean energy investment, but this remains to be seen.

To fill the tremendous energy infrastructure gap in developing and emerging countries, the public and private sectors need to work in partnership, including via Public Private Partnerships (PPPs). Such partnerships can be supported by IFIs. For large energy project partnerships, pooling public, private and donor funds should not crowd out the private sector. Instead they offer the potential to crowd in private funds to risky markets that would not receive private investment without proper risk mitigation. But PPPs need the right environment to be successful and a fair share of risks and rewards. This can be very challenging, as discussed in this book.

IFIs need to provide instruments that are more flexible and more cost effective for the private sector and with a shorter processing time. Better coordination is needed between World Bank Group institutions providing guarantees/insurances, as indeed is better marketing of those instruments. Export credit agencies can also play a constructive role in supporting exporters of equipment and services to developing and emerging markets by providing guarantees and insurance against commercial and non-commercial risks to facilitate longer term lending and at more affordable cost.

A small country like Iceland can play a constructive role in developing and emerging countries by sharing its own transition experience from fossil fuel to clean energy and could be an inspiration and an example for other countries to follow. Icelandic companies have a proven record in building and operating geothermal power plants. Several service providers, research and education institutions in geothermal energy, could also contribute to the transition to geothermal energy in developing countries. If Icelandic companies participate in energy projects as sponsors/investors in emerging market economies, they should make serious efforts to develop a comprehensive risk identification and mitigation strategy before they engage. This could be done by forming an international consortium with the participation of IFIs that Iceland is a member of and with the support of the Icelandic export credit agency, TRÚ.

The government of Iceland should increase its capacity in cooperating with those IFIs that it is a member of. It should also carry out feasibility studies that could help in the decision-making process of applying for membership in regional development banks. This could help strengthen the bargaining position of Icelandic companies *vis-à-vis* IFIs and enable them to select from a larger menu of financial and risk mitigation instruments in emerging markets than they presently can. Access to IFIs is also even more important for Icelandic companies than for companies from larger countries, as Iceland does not have wide representation in emerging markets through embassies and business representatives. The government should also investigate why Icelandic companies have not been using the services of the Icelandic export credit agency. These services could be an important export promotion tool for Icelandic exporters, especially during times of crisis.

Iceland currently provides assistance for geothermal exploration in Sub-Saharan Africa and at the same time decided to become a member of the Asian Infrastructure Investment Bank, whose concentration is in Asia. Lack of strategy, coordination and vision makes a meaningful contribution from Iceland problematic. Current efforts by individuals and companies are mostly on a small scale and fragmented. Given how small Iceland is, it is unlikely that it can alone play an important role in coordinating a global effort to promote and increase cross border clean investments internationally.

Geothermal cases discussed in this book from Indonesia, Kenya and Turkey, as well as hydropower cases from Lao and Uganda, show how successful leveraging of multilateral guarantee mechanisms can make clean energy investment possible in a challenging investment environment. National agencies such as export credit agencies also provide significant support by using their instruments to facilitate cross border trade finance. Any global climate agreement should provide incentives for all countries, especially developing countries and emerging market countries, to shift international infrastructure investment to low-carbon power from coal and oil. Here the IFIs can serve as key instruments for the global community.

This book does not argue that investment in clean energy is the only action needed to reach our climate goals. The global warming externality must be reduced by having an appropriate price for carbon. This will certainly be resisted by some well-represented fossil fuel dependent industries that seek to block policies that would benefit the larger, longer term interests of the majority of people, born and unborn. If other policies fail, however, development and investment in low-carbon technologies is the last resort for achieving our climate goals.

IFIs are well suited to mitigating non-commercial risks in emerging markets e.g., because of their global nature and funding and risk mitigation instruments they already have. In addition, they often engage in policy dialogue with host governments and can facilitate the regulatory reform often needed in emerging markets. Furthermore, they can provide the technical assistance often needed in developing and emerging markets. IFIs have also so far done little to help fund the high upfront development costs of geothermal projects in developing countries. Here they should team up with other development institutions with grants and concessional lending to help enable geothermal projects.

Local and global externalities need to be considered when assessing the viability of those projects. Low fossil fuel prices can make for fragile progress in development of renewables and can slow down both innovation and adoption of clean energy technology. Governments should not allow low fossil fuel prices to derail the transition to clean energy. They cannot afford to take that risk.

References

Arezki, R. and Obstfeld, M. (2015). The Price of Oil and the Price of Carbon, available at: http://blog-imfdirect.imf.org/2015/12/02/the-price-of-oil-and-the-price-of-carbon/ (Accessed on 7 December 2015).

ASEAN. (2014). ASEAN Principles for PPP Frameworks, available at: http://asean.org/asean-welcomes-principles-for-public-private-partnership-framework/ (Accessed on 19 June 2016).

Asian Development Bank. (2000). Review of the Partial Risk Guarantee of the Asian Development Bank, available at: http://www.adb.org/sites/default/files/institutional-document/32087/prg.pdf (Accessed on 1 June 2016).

Asian Development Bank. (2006). Review of ADB's Credit Enhancement Operations, available at: http://www.adb.org/documents/review-adbs-credit-enhancement-operations (Accessed on 9 June 2016).

Asian Infrastructure Investment Bank. (2015). Articles of Agreement, available at: http://www.aiib.org/uploadfile/2015/0814/20150814022158430.pdf (Accessed on 20 June 2016).

Cambridge Economic Policy Associates. (2015). Kenya Country Case Study - Mobilising Finance for Infrastructure - A Study for the UK Department for International Development (DFID), August 2015.

Climate Investment Funds, CIF. (n.d.). Governance, available at https://www-cif.climateinvestmentfunds.org/about/governance (Accessed on 19 November 2015).

Climate Policy Initiative. (2015a). Using Public Finance to Attract Private Investment in Geothermal: Olkaria III Case Study, Kenya, June 2015, available at: http://climatepolicyinitiative.org/wp-content/uploads/2015/06/150601_Final_Olkaria_ForWeb.pdf (Accessed on 13 November 2015).

Climate Policy Initiative. (2015b). Public Finance and Private Exploration in Geothermal: Gümüşköy Case Study, Turkey, available at: http://climatepolicyinitiative.org/publication/public-finance-and-private-exploration-in-geothermal-gumuskoy-case-study-turkey/ (Accessed on 10 November 2015).

Cuttaree, V. and Mandri-Perrott C. (2011). Public-Private Partnerships in Europe and Central Asia – Designing Crisis-Resilient Strategies and Bankable Projects. The World Bank, 2011.

Czinkota, M., Ronkainen I., Moffett M., Marinova S. and Marinov, M. (2009). International Business European Edition. England: John Wiley & Sons, Ltd.

de Palma, A., Leruth, L., Prunier, G. (2009). Towards a principal-agent based typology of risks in public-private partnerships. IMF Working Paper, WP/09/177, available at: http://papers.ssrn.com/sol3/papers.cfm?abstract_id=1475518. (Accessed 10 May 2016).

Delmon, J. (2009). Private Sector Investment in Infrastructure. Project Finance, PPP Projects and Risk. 2nd edition, Kluwer Law International, the Netherlands.

Dinh, T. Q. and Hilmarsson H. Þ. (2012a). What are the Economic Justifications for the Existence of Export Credit Agencies and How Can They Facilitate Cross Border Trade to Emerging Market Economies? Journal of Regional Formation and Development Studies, 2012, 6, p. 15-25, Klaipeda.

Dinh, T. Q. and Hilmarsson, H. Þ. (2012b). Private Sector Export to Emerging Market Economies During Times of Crisis: How Can Export Credit Agencies Help? Review of International Comparative Management, Volume 13, Issue 1, March 2012, p. 167-180.

Dinh, T. Q. and Hilmarsson. H. Þ. (2012c). How Can Private Companies Use the Financial Services and Risk Mitigation Instruments Offered by Export Credit Agencies in Emerging Markets? Proceedings. Project Development - Practice and Perspectives. First International Scientific Conference on Project Management in the Baltic Countries. February 8-9, 2012, Riga, University of Latvia, p. 14-25.

Estache, A. (2005). PPI partnerships versus PPI divorces in developing countries (or are we switching from PPPI to PPDI?). Policy Research Working Paper, 3470. Washington: World Bank, available at: http://ideas.repec.org/p/wbk/wbrwps/3470.html. (Accessed on 11 May 2016).

Gavin, M. and Rodrik, D. (1995). The World Bank in a Historical Perspective, the American Economic Review, Vol. 85, No. 2. Papers and Proceedings of the Hundredth and Seventh Annual Meeting of the American Economic Association Washington, DC, January 6 to 8, 1995, p. 329-334.

Gekon. (2011). The Icelandic Geothermal Cluster; Mapping and Mobilization, available at: http://www.icelandgeothermal.is/files/pdf/igc.mapping-and-mobilizing.2nd-september-2011.final.pdf (Accessed on 6 May 2016).

GEOELEC. (2013). European Geothermal Risk Insurance Fund, EGRIF, available at: http://www.geoelec.eu/wp-content/uploads/2011/09/D-3.2-GEOELEC-report-on-risk-insurance.pdf (Accessed on 2 March 2016).

Gore, A. (2009). *Our Choice – A Plan to Solve the Climate Crisis*. Melcher Media, New York.

Greve, C. (2015). International Public-Private Partnership Policies: Convergence in Themes from ASEAN, the European Union, IMF, OECD, the UN, and the World Bank? Cornell University, Cornell Program for Infrastructure New York City, 15-16 September 2015, available at: http://openarchive.cbs.dk/bitstream/handle/10398/9196/Greve-paper-NYC-PPP-conference-2015.pdf?sequence=1 (Accessed on 20 June 2016).

Grímsson, Ó. R. (2006). Icelandic ventures: Can the success continue? Speech made at the Kaupthing Seminar in Helsinki on May 24, 2010, available at: http://www.forseti.is/media/files/06.05.24.Helsinki.Conference.pdf (Accessed on 15 June 2014).

Grímsson, Ó. R. (2010). Iceland Geothermal 2010. Speech made in Reykjavík on November 1 during a conference entitled: Iceland geothermal 2010, available at: http://forseti.is/media/PDF/2010_11_01_Iceland_Geothermal.pdf (Accessed on 5 November 2015).

Hart, O. (2003). Incomplete contracts and public ownership: Remarks, and an application to public-private partnerships. Economic Journal, 113, 69–76, available at: http://scholar.harvard.edu/files/hart/files/incompletecontractsandpublicownershipej.pdf (Accessed on 5 May 2016).

Herrera-Martínez, A. (2014). EBRD's Geothermal Experience: Combining financing with technical assistance and policy dialogue under the Sustainable Resource Initiative. *Energy Efficiency and Climate Change*, E2C2.

Herrera-Martínez, A. (2015). The Role of the Private Sector in the Development of Geothermal Power – EBRD's Experience. Energy Efficiency and Climate Change, EBRD, Kanyon Ofis, Istanbul, Turkey, available at: http://climatepolicyinitiative.org/wp-content/uploads/2015/07/HERRERA-MART%C3%ACNEZ-The-Role-of-the-Private-Sector-in-the-Development-of-Geothermal-Power-%E2%80%93-EBRD%E2%80%99s-Experience.pdf (Accessed on 10 June 2016).

Hilmarsson, H. Þ. and Dinh T. Q. (2013). Export credit agencies, cross border trade, times of crisis, and increased value added in emerging economies' industrial sectors. Conference article presented at the 55th Annual Meeting of the Academy of International Business, 3 to 6 July, 2013, Istanbul, Turkey.

Hodge, G. A. and Greve, C. (2009). PPPs: *The Passage of Time Permits a Sober Reflection.* Institute of Economic Affairs, Blackwell Publishing, Oxford, Vol. 29. Issue 1, 33-39.

Iceland Geothermal. (2016). Role of Iceland Geothermal, available at: http://www.icelandgeothermal.is/iceland-geothermal/role/ (Accessed on 7 June 2016).

Icelandic International Development Agency. (2014). Geothermal, available at: http://arsskyrslur.iceida.is/en/2014/regional-offices/geothermal/ (Accessed on 7 June 2016).

International Financial Corporation, IFC. (2009). Financing of Renewable Energy: IFC's Perspective. Presentation made in Washington D.C. in November 2009. Washington: World Bank.

International Financial Corporation, IFC. (2011). International Finance Institutions and Development Through the Private Sector, available at: http://www.miga.org/documents/IFI_report_09-13-11.pdf (Accessed on 15 June 2016).

International Financial Corporation, IFC. (2014). Handshake: IFC's quarterly journal on public-private partnerships, available at: http://documents.worldbank.org/curated/en/2014/04/19456522/handshake-ifcs-quarterly-journal-public-private-partnerships-13 (Accessed on 16 October 2015).

International Monetary Fund. (2004). Public Private Partnerships. International Monetary Fund, Washington, DC, available at: http://www.imf.org/external/np/fad/2004/pifp/eng/031204.pdf (Accessed on 15 June 2016).

Keohane, R.O. and Victor, D. G. (2011). The Regime Complex from Climate Change. Perspectives on Politics, March 2011, Vol. 9/No. 1, pp 7 – 23.

Keohane, R.O. (2015). *The Global Politics of Climate Change: Challenge for Political Science,* The 2014 James Madison Lecture, PS, January 2015.

Ketels, C. and Memedovic, O. (2008). From clusters to cluster-based economic development. *International Journal of Technological Learning, Innovation and Development*, 1, (3), 375-392.

Ketels, C. (2010). Profile of the Icelandic Geothermal Cluster. Presentation made at Icelandic Geothermal 2010. Conference in Reykjavík on 1 November 2010.

Krueger, A. O. (1997). Whither the World Bank and the IMF? Working Paper 6327. National Bureau of Economic Research, Cambridge MA, available at: http://www.nber.org/papers/w6327 (Accessed on 15 June 2016).

Leruth, L. E. (2009). Public-Private Cooperation in Infrastructure Development: A Principal-Agent Story of Contingent Liabilities, Fiscal Risks, and Other (Un)pleasant Surprises. Springer Science + Business Media, LLC, available at: http://link.springer.com/article/10.1007%2Fs11067-009-9112-0 (Accessed on 15 June 2016).

Morse, J. C., and Keohane, R. O. (2014). Contested multilateralism. The Review of international organizations, 9(4), 385-412, available at: http://www.princeton.edu/~rkeohane/publications/PublishedCM.pdf (Accessed on 15 June 2016).

New Development Bank (NDB) - BRICS. (2014). Agreement on the New Development Bank, July 15, 2014, Fortaleza, Brazil, available at: http://ndbbrics.org/agreement.html (Accessed on 1 June 2016).

Nordhaus, W. (2013). *The Climate Casino – Risk, Uncertainty, and Economics for a Warming World.* Yale University Press, New Haven & London.

Organization for Economic Cooperation and Development. (2012). Recommendations of the Council on Principles for Public Governance of Public-Private Partnerships. Paris: OECD, available at: https://www.oecd.org/governance/budgeting/PPP-Recommendation.pdf (Accessed on 20 July 2015).

Porter, M. E. (1998). Clusters and the new economics of competition. *Harvard Business Review.* November-December 1998.

Porter, M. E. (2000). Location, competition, and economic development: Local clusters in a global economy. *Economic Development Quarterly*, 14(1), 15-34.

Porter, M. E. (2010). The Icelandic Geothermal Cluster: Enhancing Competitiveness and Creating a new Engine of Icelandic Growth. Presentation made at Icelandic Geothermal 2010. Conference in Reykjavík on 1 November 2010.

Porter, M. E. (2016). Unlocking the Potential of Geothermal Energy: Strategic Implications for Iceland, Other Nations and the International Community. Presentation given at the Iceland Geothermal Conference, 28 April 2016.

Porter, I. C. and Shivakumar, J. (2010). Doing a dam better: the Lao People's Democratic Republic and the story of Nam Theun 2 (NT2). Washington D.C. - The World Bank, available at: http://documents.worldbank.org/curated/en/2010/01/13240425/doing-dam-better-lao-peoples-democratic-republic-story-nam-theun-2-nt2. (Accessed on 2 June 2016).

Rhodes, F.H.T. (2012). *Earth – A Tenant's Manual.* Cornell University Press, Ithaca and London.

Salacuse, J. W. (2010). The Emerging Global Regime for Investments, Harvard International Law Journal, Volume 51, Number 2, Summer 2010, available at: http://www.harvardilj.org/wp-content/uploads/2010/09/HILJ_51-2_Salacuse.pdf (Accessed on 2 June 2016).

Stern Review. (2006). Stern Review: The Economics of Climate Change, available at: http://mudancasclimaticas.cptec.inpe.br/~rmclima/pdfs/destaques/sternreview_report_complete.pdf (Accessed on 2 August 2015).

Stiglitz, J. (2009). G7 Takes Back Seat as Crisis Pushes G20 to Fore. In Sunday Morning Herald, available at:http://www.smh.com.au/business/world-business/g7-takes-back-seat-as-crisis-pushes-g20-to-fore-20090213-8705.html (Accessed on 20 June 2015).

Wells, L. T. (2005). The New International Property Rights. Can the Foreign Investor Rely on Them? In International Political Risk Management. Looking to the Future, Theodore H. Moran and Gerald T. West, Eds. The World Bank Group – Multilateral Investment Guarantee Agency.

Wells, L.T. and Ahmed, R. (2007). *Making Foreign Investment Safe – Property Rights and National Sovereignty*. Oxford University Press.

West, G. T. (1999). Political Risk Investment Insurance: A Renaissance. *Journal of Project Finance*, 5(2), 27-36.

Wolfensohn, J.D. (2010). A Global Life. My Journey among Rich and Poor, from Sydney to Wall Street to the World Bank. Public Affairs, New York.

World Bank. (1994). *Mainstreaming of Guarantees as an Operational Tool of the World Bank*. Washington D.C.: Author, R94-145.

World Bank. (2003). Global Economic Prospects and the Developing Countries in 2003.

World Bank. (2005). Project Finance and Guarantees – IDA Guarantee Paves Renewed Interest in Private Hydropower – Nam Theun 2 Project, available at: http://siteresources.worldbank.org/INTGUARANTEES/Resources/Lao_NamTheun2_Note.pdf (Accessed on 8 June 2016).

World Bank. (2007). Financial Solutions: Partial Risk Guarantee. IDA Partial Risk Guarantee (PRG) to promote Uganda's first Independent Power Producers, available at: http://siteresources.worldbank.org/INTGUARANTEES/Resources/UgandaBujagaliNew.pdf (Accessed on 8 May 2016).

World Bank. (2008). Capital Matters, Vietnam Consultative Group Meeting Hanoi, 2008, available at: http://siteresources.worldbank.org/ INTVIETNAM/Resources/387318-1235543674588/vn_devreport09_ front.pdf (Accessed on 7 June 2016).

World Bank. (2009). The World Bank Group Guarantee Instruments 1990-2007. An Independent Evaluation. The World Bank Washington, D.C., available at: http://siteresources.worldbank.org/EXTGUARANTE/ Resources/guarantees_eval_full.pdf (Accessed on 2 June 2016).

World Bank. (2010). World Development Indicators Database, available at: http://siteresources.worldbank.org/DATASTATISTICS/Resources/GNI.p df(Accessed on 2 June 2016).

World Bank. (2012a). Green Infrastructure Finance – Framework Report. East Asia and Pacific Region/East Asia Infrastructure Unit (EASIN).

World Bank. (2012b). Geothermal Handbook: Planning and Financing Power Generation. Washington DC: World Bank Group, Energy Sector Management Assistance Program, available at:http://www.esmap.org/sites/ esmap.org/files/DocumentLibrary/FINAL_Geothermal%20Handbook_TR 002-12_Reduced.pdf (Accessed on 19 June 2014).

World Bank. (2013a). Doing Business 2014 - Understanding Regulations for Small and Medium-Size Enterprises, available at: http://doingbusiness.org/ ~/media/GIAWB/Doing%20Business/Documents/Annual-Reports/ English/DB14-Full-Report.pdf (Accessed on 2 June 2014).

World Bank. (2013b). Global Geothermal Development Plan, available at: http://www.esmap.org/sites/esmap.org/files/ESMAP_Paris_Geothermal_E nergy_KEF_Optimized.pdf (Accessed on 12 June 2015).

World Bank. (2013c). Enhancing the World Bank's Operational Policy Framework on Guarantees. Operations Policy and Country Services November 19, 2013, available at: http://documents.worldbank.org/curated/ en/2013/11/18542542/enhancing-world-banks-operational-policy-framework-guarantees (Accessed on 12 October 2015).

World Bank. (2013d). Project Appraisal Document on a Proposed IDA Credit in the Amount of SDR4 Million (US$6 million Equivalent) and a Proposed Grant From the Global Environment Facility Trust Fund in the Amount of US$6.04 Million and a Proposed Grant from ESMAP in the Amount of US$1.10 Million to the Republic of Djibouti for a Geothermal Power Generation Project, available at: http://www-wds.worldbank.org/ external/default/WDSContentServer/WDSP/IB/2013/05/20/000442464_2 0130520095050/Rendered/PDF/682540PAD0P127010Box377300B00OU O090.pdf (Accessed on 12 December 2015).

World Bank. (2014). Scaling-up Renewable Geothermal Energy in Indonesia – An Integrated Approach to Evaluating a Green Finance Investment, available at: https://www.esmap.org/sites/esmap.org/files/Document Library/ESMAP_Scaling-up%20Geothermal%20in%20Indonesia_KS15-13_Optimized.pdf (Accessed on 15 June 2016).

World Bank. (2015a). The World Bank Group A to Z. The International Bank for Reconstruction and Development/The World Bank, Washington, DC 20433.

World Bank. (2015b). An Update on the Global Geothermal Development Plan De-Risking Exploration to Scale-up Geothermal Energy, ESMAP- World Geothermal Congress 2015, Donors Forum.

World Bank. (2015c). Kenya: Olkaria III Geothermal. Public-Private Partnerships Briefs May 2015, available at: http://www-wds.worldbank.org/external/default/WDSContentServer/WDSP/IB/2015/06/04/090224b082ef31c9/1_0/Rendered/PDF/Kenya000Olkaria0III0geothermal.pdf. (Accessed on 20 June 2016).

World Economic Forum. (2006). Building on the Monterrey Consensus: The Untapped Potential of Development Finance Institutions to Catalyze Private Investment. World Economic Forum Financing for Development Initiative.

Zoellick, R. B. (2009). Seizing Opportunity from Crisis: Making Multilateralism Work, available at: http://web.worldbank.org/WBSITE/EXTERNAL/EXTABOUTUS/ORGANIZATION/EXTPRESIDENT/EXTPASTPRESIDENTS/PRESIDENTEXTERNAL/0,,contentMDK:22121476~menuPK:232084~pagePK:159837~piPK:159808~theSitePK:227585,00.html (Accessed on 6 November 2015).

ANNEX

Regional IFIs and their financial instruments
The **African Development Bank**[99] continues to adjust its operations by adapting existing instruments and adopting new instruments, when necessary, to manage risk and to assure efficient delivery of services to the changing needs of its clients in emerging markets, inter alia, the following financial products: • **Senior Debt**: – Up to 20 years (five years' grace period). – Foreign or local currencies, risk priced. • **Guarantees**: – Partial risk, partial credit (long tenors). – For specific credit or commercial bank small and medium-sized enterprise programs, risk priced. • **Subordinated Debt**: – Subordinated loans or bonds. – Local currency possible, maximum 15 years (10+5), risk priced. • **Equity**: – Direct investments in banks, development financial institutions, micro-finance institutions, etc. – Up to 25 per cent with or without board seat, planned exit, commercial returns. • **Technical Assistance**: – Grants for studies.

99 See further: http://www.afdb.org/en/topics-and-sectors/sectors/private-sector/types-of-financial-instruments/.

– Up to US $1 million for institutional capacity building programs.
The **Asian Development Bank** [100] catalyzes private investments through direct financing, credit enhancements, and risk mitigation instruments. The Asian Development Bank provides direct funding assistance through loans and equity investments and offers political risk guarantee and partial credit guarantee instruments to enhance the risk profiles of transactions to attract both foreign and local commercial lenders to projects in the developing member countries, and to encourage them. Through co-financing and guarantees, the Asian Development Bank also supports local investors, domestic banks, and financial institutions to provide funds on suitable terms for Asian Development Bank-assisted development projects. The Asian Development Bank also mobilizes additional resources for projects through a "B-loan" arrangement. • **Loans** The Asian Development Bank offers hard currency loans, both senior and subordinated, as well as mezzanine financing. It also offers local currency loans in selective markets on a case to case basis. Interest rates and other terms vary, depending on a company or project's needs and risks. • **Equity investments** The Asian Development Bank may invest directly in an enterprise. • **Guarantees** The Asian Development Bank provides guarantees that can be used to cover political risks, and more comprehensively, both political and credit risks. • **B-loan/Lender of record** The Asian Development Bank can mobilize additional debt resources for financing through an arrangement in the form of a B-loan/Lender of Record structure. • **Technical assistance** In addition to financial products, the Asian Development Bank also offers technical assistance on a selective basis, for public and private sector operations.
The **European Bank for Reconstruction and Development** (EBRD)[101] offers financial products tailored to each client. Prospective clients have to demonstrate that their proposed project or business meets the minimum

[100] See further: http://www.adb.org/site/private-sector-financing/financial-products.

[101] See further: http://www.ebrd.com/what-we-do/products-and-services.html.

requirements to be eligible to be considered for EBRD involvement. The forms of direct financing are:

- **Loans**
- **Equity Investments**
- **Guarantees to promote trade**

Policy Reform Dialogue services

At the EBRD the Policy and Partnerships department is responsible for leading the coordination of strategies, policies and related initiatives supporting the Bank's transition mandate.

Advisory services

The EBRD's Small Business Support team helps small and medium-sized businesses grow, succeed, then grow again, becoming genuine catalysts for their local economies and region.

Inter-American Development Bank[102]

Non-Sovereign Financing

For its non-sovereign guaranteed borrowers, the Inter-American Development Bank provides private and state-owned companies and financial institutions in Latin America and the Caribbean with the financing necessary to help meet the region's growing demands.

Types of Financing

Financing is offered via a variety of structures including loans, partial credit guarantees, grant financing and political risk guarantees. Loans are made on a project finance or on a corporate finance basis. Partial credit guarantees help improve placement of bonds with investors and enable clients to have access to better financing terms from domestic and foreign investors. Grant financing is available to private and state owned enterprises for pre-investment activities of selected highly developmental projects. Political risk guarantees are provided to eligible foreign debt investors for qualified investments in borrowing member countries.

[102] See further: //www.iadb.org/en/idb-finance/english/non-sovereign-financing,1985.html.

General financing terms

- **Loans** carry fixed or floating interest rates and are priced according to existing market conditions, reflecting the overall characteristics of the underlying project and client.
- **Partial credit guarantees** carry fees or premiums that are market based and depend on the risk covered and market benchmarks.
- **Grant** financing requires counterpart financing.
- **Political risk guarantees** carry fees determined on a case-by-case basis depending on the risk covered and the nature of the guarantees' structure.

European Investment Bank (EIB)[103]

Lending is by far the EIB's principal activity, accounting for around 90% of total financial commitment. The EIB lends to clients of all sizes to support sustainable growth and jobs. EIB support can be central to attracting other investors.

The EIB holds a variety of tools to help clients blend EIB financing with additional sources of investment. The EIB also offers advice to help maximize value for money.

- **Lending**
- **Blending**
- **Advising**

[103] See further: http://www.eib.org/products/index.htm.

BOOK ENDORSEMENTS

Developing and emerging economies are rapidly approaching the technological frontier through various structural changes, and huge increases in their energy requirements are foreseeable. In his new book, Professor Hilmarsson is concerned with ways to minimize the likely environmental damage and offers sensible ideas about ways to encourage a large-scale transition to clean energy sources, in particular how best to stimulate investments in geothermal and hydropower projects. He makes a well-informed and convincing case that international financial institutions, including the World Bank Group, in partnership with the private sector, should scale up their efforts and co-ordinate investment in clean energy infrastructures. In spite of alarming reports, the practical response of international financial institutions to the environmental crisis is lukewarm. Hilmarsson makes a valuable contribution by mapping the way ahead.

Reviewed by Professor Thráinn Eggertsson, Institute of Economic Studies, University of Iceland and Hertie School of Governance, Berlin, Germany

In this book Professor Hilmarsson, a World Bank Group insider for 12 years, offers an interesting and well-informed analysis of international financial institutions (IFIs) and how they can help solve one of the world's greatest problems - the transition to clean energy. Too often IFIs' efforts are hampered by bureaucratic procedures that make them less effective and their clean energy portfolio in developing and emerging countries still remains small. In addition to funding and risk mitigation instruments, IFIs are particularly well placed to provide policy advice and technical assistance to governments implementing energy sector reform. This book is an important and timely contribution to the debate on clean energy and climate change.

Reviewed by Professor Dr. Erika Sumilo, Head of the Department of International Economics and Business, University of Latvia

The utilization of clean energy resources in developing and emerging countries is among key challenges humankind is currently facing. Professor Hilmarsson offers a well-informed analysis on how international financial institutions can contribute more effectively to the transition to clean energy -- a critical component for success in the battle against climate change. This book is insightful and the analysis is sound as Professor Hilmarsson draws on his experience from the World Bank for 12 years in three different continents.

Reviewed by Professor Ligita Šimanskienė, Klaipėda University, Lithuania

Clean energy investments, including geothermal and hydropower projects, tend to be large, with long repayment periods and often require cooperation between public, private and donor partners in emerging countries. Professor Hilmarsson shows how the funding and risk mitigation products of international financial institutions can be instrumental in mobilizing funds to make clean energy investments both possible and feasible. This is an important topic as vast clean energy sources still remain underutilized in emerging countries during a time of growing energy demand. Professor Hilmarsson also discusses Iceland's transition to clean energy from which other countries may draw some valuable lessons. This is an important contribution that deserves attention and debate.

Reviewed by Professor Anatoliy G. Goncharuk, Head of the Department of Business Administration and Corporate Security, International Humanitarian University, Ukraine

As Professor Hilmarsson highlights in this book, no institution supports cross border investments in a similar way that the World Trade Organization does for cross border trade. This failure traces its roots all the way back to the Bretton Woods conference and is an impediment for clean energy investments, especially in developing and emerging countries, during a time of great need for a transition to clean energy sources. In this book Professor Hilmarsson shows how international financial institutions can more effectively support clean energy investments in a world challenged by climate change. A timely contribution and an enjoyable read.

Reviewed by Dr. Sujit Kumar, Research Scholar, Jawaharlal Nehru University, New Delhi, and Assistant Professor, Sri Venkateswara College, University of Delhi, India.

About the Author

Hilmar Þór Hilmarsson is a Professor at the University of Akureyri, School of Business and Science, Iceland. Dr. Hilmarsson teaches courses on international business and macroeconomics. He received his cand. oecon. degree in Economics from the University of Iceland in 1987, an MA in Economics from New York University in 1989, and a Ph.D. in public administration and economic development from the American University in Washington, D.C. in 1992. He served as a Specialist and Coordinator with the World Bank Group in Washington D.C. from 1990 to 1995, at the World Bank office in Riga from 1999 to 2003 and the World Bank office in Hanoi from 2003 to 2006. From 1995 to 1999 he served as a Special Advisor to the Minister for Foreign Affairs in Iceland. Dr. Hilmarsson has published over 50 scholarly articles and book chapters and has lectured and made presentations in more than 30 universities in Europe and the USA, including the American University in Washington DC, Aalborg University, Cornell University, Georgetown University, Klaipeda University, Stockholm School of Economics, University of California Berkeley, University of California Los Angeles, the University of Latvia, the University of Mauritius, the University of Porto, the University of Tartu, the University of Washington, Vytautas Magnus University, the University of York, and Yale University. He has travelled to about 60 countries. He was a Visiting Professor at Stockholm School of Economics in Riga during the fall semester 2013, a Visiting Scholar at the University of Washington in Seattle in the spring semester 2014 and a Visiting Professor at Cornell University during the fall semester 2015. Dr. Hilmarsson is a Visiting Scholar at the University of California, Berkeley during the fall semester 2016. He can be contacted at Hilmar@unak.is.

INDEX

D

E

H

I

K

L

M

N

S

T

U

V

W

Z